PAINT AND COATINGS INDUSTRY

Edited by **Faris Yilmaz**

Paint and Coatings Industry
http://dx.doi.org/10.5772/intechopen.71971
Edited by Faris Yilmaz

Contributors

Imed Gargouri, Khadhraoui Moncef, Leon Begrambekov, Andrey Grunin, Nikolai Puntakov, Yaroslav Sadovskiy, Vyacheslav Budaev, Sergey Grashin, Jerzy Zych, Jan Mocek, Alfonso Martone, Stefania Dello Iacono, Eugenio Amendola, Chunfu Chen, Zoi Manoli, Darja Pecko, Guy Van Assche, Herman Terryn, Johan Stiens, Ali Pourkazemi, Pierre Verge, Acerina Trejo-Machin, Laura Puchot

First published in London, United Kingdom, 2019 by IntechOpen
IntechOpen is the global imprint of INTECHOPEN LIMITED, registered in England and Wales, registration number: 11086078, The Shard, 25th floor, 32 London Bridge Street
London, SE19SG – United Kingdom
Printed in Croatia

British Library Cataloguing-in-Publication Data
A catalogue record for this book is available from the British Library

Additional hard copies can be obtained from orders@intechopen.com

Paint and Coatings Industry, Edited by Faris Yilmaz
p. cm.
Print ISBN 978-1-78985-161-8
Online ISBN 978-1-78985-162-5

For EU product safety concerns:
IN TECH d.o.o., Prolaz Marije Krucifikse Kozulić 3, 51000 Rijeka, Croatia,
info@intechopen.com or visit our website at intechopen.com.

Meet the editor

Dr. Faris Yılmaz was born in Palestine on October 7, 1969. After precollege education at Jinin-West Bank, he was accepted into the Chemical Engineering Department of the Middle East Technical University in Ankara, Turkey. There he obtained his BSc degree in Chemical Engineering and his MS and PhD degrees in Polymer Science and Technology. His research has mainly concentrated on conducting polymers, their solution properties, composites, and nanocomposites. He has publications in different polymer journals. He is married and has two children.

Contents

Preface

Pigmented liquids, liquefiables, or mastic compositions that after application to a substrate in a thin layer converts to a solid film are called *paints* or *coatings*. They are used to protect, color, or provide texture to objects. They can be made or purchased in many colors —and in many different types, such as watercolor, synthetic, etc. They are also stored, sold, and applied as a liquid, but most types dry to a solid.

In 1718, Marshall Smith invented a "machine or engine for the grinding of colours" in England. It is not known precisely how it operated, but it was a device that increased the efficiency of pigment grinding dramatically. Soon, a company called Emerton and Manby was advertising exceptionally low-priced paints that had been ground with labor-saving technology.

By the proper onset of the industrial revolution, paint was being ground in steam-powered mills and an alternative to lead-based pigments was found in a white derivative of zinc oxide. Interior house painting increasingly became the norm as the 19th century progressed, both for decorative reasons and because the paint was effective in preventing the walls from rotting from damp. Linseed oil was also increasingly used as an inexpensive binder.

The maximal changes that have taken place in the coatings industry during the last 50 years were the espousal of new coating technologies, including waterborne (thermosetting emulsion, colloidal dispersion, water-soluble) coatings, high-solids coatings, two-component systems, powder coatings, and radiation-curable coatings.

There are two primary functions provided by coatings—decoration and protection—that are of substantial economic importance. Forty-five percent of the coatings produced worldwide are used to decorate and protect new constructions as well as to maintain existing structures, including residential homes and apartments, public buildings, and plants and factories. Another 40% of the coatings are used to decorate and/or protect industrial products. Without coatings, product life can be significantly shortened and many products would not even be marketable. Most of the remaining coatings, called "special purpose," are used for various applications such as traffic paints, vehicle refinishing, high-performance coatings for industrial plants and equipment, and protection of marine structures and vessels. These are usually applied in outdoor conditions.

Through the coming years, air pollution regulations will continue to be a driving force behind the adoption of new coating technologies. Despite the overall relatively slow growth in demand anticipated for coatings, waterborne and high-solids coatings, powders, UV curables, and two-component systems seem to have good growth potential.

Nanotechnology is one new area of interest with tens of thousands of patents currently only available for the paint industry. Very small ceramic or metallic particles can be added to paint formulations to modify specific properties in highly specialized applications. The average size of nanoparticles is 10–70 nm, consisting of less than 6.5 million atoms. At these sizes, the ratio of surface area to mass becomes significant, giving the particles unique properties.

In this book an effort has been made to achieve justice for various topics and research related to the paint and coatings industries. In the first chapter, the discussion centers on the metrology of organic solvents in the shoe industry to Sfax city (Tunisia). Thermal volumetric analysis—a new test method of the kinetics of gas emissions from molding sands and protective coatings heated by liquid alloy—is illustrated in the second chapter. The third chaper discusses Diels–Alder chemistry to develop self-healing epoxy resins and their composites. Furthermore, design and synthesis of biobased benzoxazines is discussed in the fourth chapter. UV and thermal cure epoxy adhesives, transport of electrolyte in organic coatings on metal, and boron carbide coating behavior under high plasma heat are debated in the fifth, sixth, and seventh chapters, respectively.

Finally, I would like to take this opportunity to express my deepest thanks to my sweetheart wife Adv. Saada Idris Yilmaz for the infinite support and encouragement she has given me, which helped me complete this work.

Dr. Faris Yilmaz
Nurol Makina A.Ş.
Ankara, Turkey

Chapter 1

Metrology Organic Solvents in the Shoes Industry in Sfax City (Tunisia)

Imed Gargouri and Moncef Khadhraoui

Additional information is available at the end of the chapter

http://dx.doi.org/10.5772/intechopen.78316

Abstract

In spite of the importance to the use of organic solvents in the shoes industry, exposure data of the employees at these solvents in this sector are absent in Tunisia. The objective of this study is to establish a biotoxicological supervision of exposure in shoes manufacturing companies. After the inventory of the most dominant solvents (acetone, cyclohexane, hexane and methylethylketone, toluene) in the preparations used in shoes manufacturing, 18 voluntary companies benefited from 55 dynamic atmospheric samplings realized on a duration of 4 hours. The exposure index some mixture ranged between 0.1 and 8.8 and presented an average value superior to 1, mainly in the partly industrialized process. These values reached, respectively, 1.7, 2.5 and 4.5 for the posts of finishing, shoes collector and glues dispatcher. The atmospheric average concentrations of certain solvents exceeded the limit value of professional exposure mainly for the hexane with a value of 214 mg/m^3 in certain posts. The chronic exposure to organic solvents in a shoes industry and to establish a first report on the profiles of exposure to these solvents in this sector. Hence, an approach of evaluation of the professional risk by the biotoxicological supervision of exposure is established in Sfax city.

Keywords: solvents, shoes, evaluation of the risk, metrology of atmosphere

1. Introduction

Organic solvents constitute a set of varied chemicals whose essential chemical properties are a solubilizing power associated with a generally high-volatility [1]. These properties make them products of use that are difficult to circumvent in various industrial sectors such as footwear manufacturing [2–4]. Among the chemical agents identified in this sector in Tunisia, organic solvents play a major role in the quantities used and the number of employees exposed.

However, employee exposure data for organic solvents in this sector are almost absent. The prevention of occupational risks (POR) and more particularly the chemical risk is based on the risk assessment (RA) according to procedures established by regulatory texts [4–6]. However, the Tunisian occupational health regulations [7, 8], did not provide any obligation for companies to perform RA requiring metrology of work environments, and biotoxicological dosages. The study objective is to establish a biotoxicological exposure metrology for solvent mixtures, composed mainly of acetone, cyclohexane, n-hexane, methyl ethyl ketone (MEK), and toluene used in the shoes industry of a pilot study to generalize it in all industrial sectors of Sfax city.

2. Methods

2.1. Manufacturing processes for shoes and workstations exposed to solvents

The shoe industry, despite the mechanization, remains a labor industry with 150 operations needed to make a pair of shoes. Three types of manufacturing techniques exist: welded, Goodyear sewn, and direct injected [9].

We limited ourselves in our study: (i) to the "welded" manufacturing process. It is the most widespread technique in Sfax city, and the most exposed to glue and solvent preparations (after creation, modeling, patronage, cutting and stitching, welded is the phase during, which the sole is fixed by gluing to a rod mounted on the base) and (ii) the three main positions exposing the solvents, which are the tiller station (mounting and gluing), the foundry station (display and welding), and the finishing station (cleaning, potting and packing) [10].

2.2. Sampling of shoe companies

Shoe manufacturing companies in the Sfax region have been classified and classified in three groups according to their manufacturing processes: industrial (26 companies and 751 employees), semi-industrial (6 companies and 46 employees) and handicraft (60 companies and 350 employees). The artisanal process has been divided into two categories: type 1 (foundry post and separate tiller station) and type 2 (the two spots are made by the same person in two steps).

Twenty-two companies using the "welded" process: 6 industrial, 6 semi-industrial and 10 artisanal with a workforce of 122, 48 and 60 employees (230 employees in total) were selected for this biometric study.

2.3. Selection of solvents to measure

Following a tracking of the solvents present in the composition of the products handled (glues, thinners and paint removers) in the shoe manufacturing companies during the first half of 2008, according to the simplified method of the National Institute for Research and Security (INRS) chemical RA, which mainly includes three phases: (i) inventory of products and materials used in the facility, in a workshop or workstation, (ii) prioritization of potential risks and (iii) RA [4]; we have identified five predominant solvents: acetone, cyclohexane, hexane, methyl ethyl ketone and toluene [4, 11–14].

2.4. Methods

The exposure evaluation to these 5 solvents was carried out over 15 weeks from May 20 to September 15, 2010, divided into two periods, each comprising two components [4, 5, 15, 16]: (i) an observation of the different workstations with the help of the head of the company and the oldest employees and (ii) ambient metrology of solvents and glues.

2.4.1. Biometric measurements

In order to measure the concentration of solvents in the atmosphere, individual and/or stationary samples were taken (**Figure 1**): the sensors were placed close to the airways (personal sampling system) or installed at medium height of the tracks. Above ground level (stationary system), and consisted of a sampling pump (Poket Pump: Pump SKC® 210–1002 TX) with a regular flow rate of 100 (± 5%) cm^3/min and an activated carbon trapping tube (SKC® Tube 226-16: 800 and 200 mg). The sampling was set up for a half-workstation or duration of 4 hours. The samples were taken in the middle of the week: Wednesday or Thursday depending on the type of manufacturing process of the companies and their places of installation. Solvent concentrations were determined by gas chromatography after desorption of sampled sample tubes [17–21].

(a) (b)

Figure 1. Atmospheric metrology in a semi-industrial shoe manufacturing company. (a) Individual sampling at the "uppers preparation workstation and (b) ambient sampling at the "finishing workstation.

Solvent	N° CAS[1]	France *UE*[2]		USA *ACGIH*[3]		Allemagne *MAK*[4]		Sfax study *Values selected*	
		VME[5]		TLV-TWA[6]				AEV adopted	
		ppm[7]	mg.m^{-3}	ppm	mg.m^{-3}	ppm	mg.m^{-3}	ppm	mg.m^{-3}
Acetone	67–64-1	500	1210	500	—	500	1200	500	1200
Cyclohexane	110–82-7	200	700	300	—	200	700	200	700
n-Hexane	110–54-3	20	72	50	—	50	180	20	72
Methylethylketone (MEK)	78–93-3	200	600	200	—	200	600	200	600
Toluene	108–88-3	50	192	50	—	50	190	50	190

[1]CAS: Chemical Abstract Service.

[2]UE: Union Européenne.

[3]ACGIH: American Conference of Governmental Industrial Hygienists.

[4]MAK: Maximale Arbeitsplatz-konzentration.

[5]VME: Average Exaposure Value (Valeur Moyenne d'Exposition) calculated in relation to a reference period of 8 working hours per day and 35 hours per week.

[6]TLV-TWA: Time-Werghted Average (Valeurs Weighted average values over 8 hours per day and 40 hours per week).

[7]ppm: parts-per million and volume of air.

Table 1. Occupational exposure limit values [7, 8, 11, 15, 23, 25, 26].

2.4.2. Biometric exposure limit values

The measurements results were compared with the 's (AEV) of solvents calculated for a reference period of 8 working hours/day and 39 hours/week (**Table 1**) [7, 9, 11, 22, 23]. In the absence of Tunisian exposure limit values (ELV) [9, 10], we have referred to international values (French, American and German), adopting the most restrictive values (**Table 1**) [11, 22, 24–26, 27].

Since glues are a mixture of solvents, we have conventionally used an exposure index (I.exp) for individual samples that are used as a comparison with limit values [11, 28]. This I.exp. is equal to $C_1/AEV_1 + C_2/AEV_2 + \ldots\ldots\ldots\ldots + C_n/AEV_n$ (C_n and AEV_n being, respectively, the concentration and the limit value of the pollutant n). If I.exp. is greater than 1, the limit value is considered exceeded. Calculated from a fixed-rate levy, this index is called pollution index (I.pol).

3. Results

3.1. Participating shoe companies

Eighteen companies out of the 22 companies contacted agreed to participate (participation rate of 81.8%), including four manufacturing shoes using an industrial process, five using a semi-industrial process, and nine using an artisanal process. They were divided into two groups according to the level of activity and the planning of the realization of the interventions: the first period, from May 28 to July 21, 2010, was a period of average activity (end of the preparations of the collection of summer); and the second period, from August 27 to September 12, 2010, a period of significant activity (preparations for the winter collection, the return to school, and the holiday season after Ramadan). A total of 55 atmospheric samplings of which 33 individual and 22 fixed samplings: 17 industrial enterprises, 23 semi-industrial enterprises, and 15 artisanal enterprises.

3.2. Atmospheric dosages of solvents

Employee exposures according to the industrial process show high levels of I.exp., especially in semi-industrial and artisanal type 1 enterprises during the two periods of activity, particularly during the period of high-activity (**Table 2**).

The atmospheric concentrations according to the workstation shows I.exp. and/or I.pol exceedances for the exhibitor stations (tigers, foundries, and finishes) in the processes: industrial, semi-industrial and artisanal type 1 (**Table 3**). At the silkscreen station in the industrial process, occupied by a young woman of childbearing age, the I.exp. is equal to 9.4.

The set of atmospheric measurements carried out in the different companies shows that the solvent exposure varies from one process to another and from one station to another (**Table 4**). The average atmospheric concentrations of hexane were particularly high with exceedances of the AEV including the position of the tiller, the melter, and the finish in all processes with the exception of the artisanal process type 2. With the exception of MEK at the uppers preparation

Process	Company	Samples						
		Number	Individual			Ambience		
			Exposure index (I.exp)			Pollution index (I.pol)		
			n	min-max	average	n	min-max	average
Industrial	1	5	3	0.3–1.1	0.7	2	0.5–0.8	0.6
	2	2	1	–	**1.8**	1	–	0.3
	3	5	4	0.9–9.4	**3.7**	1	–	0.8
	4	5	4	0.8–19.8	**6.8**	1	–	0.0
Semi-Industrial	1	5	3	0.4–1.2	0.7	2	0.0–0.3	0.1
	2	5	4	0.5–2.0	1.0	1	–	0.4
	3	4	2	0.2–2.5	**1.3**	2	0.2–0.4	0.3
	4	5	4	0.9–2.4	**1.7**	1	–	0.1
	5	4	1	–	**5.6**	3	4.3–8.8	**6.8**

Process		Company	Samples						
			Number	Individual			Ambience		
				Exposure index (I.exp)			Pollution index (I.pol)		
				n	min-max	average	n	min-max	average
Artisanal	Type 1	1	2	1	—	0.1	1	—	0.4
		2	1	/			1	—	0.7
		3	1	/			1	—	0.8
		4	3	2	0.4–1.7	1.1	1	—	**1.6**
		5	2	1	—	**1.5**	1	—	0.9
		6	2	2	1.6–3.4	**2.5**	/		
		7	2	1	—	**6.4**	1	—	**5.1**
	Type 2	8	1	/			1	—	0.5
		9	1	/			1	—	0.6
				Total = 33			**Total = 22**		

First Period: May 28 to July 21, 2008.
Second Period: from August 27 to September 12, 2008.
n: number of measurements taken.

Table 2. Atmospheric exposure of employees according to the manufacturing process in the 18 companies.

Process	Workplace	Samples						
		Number	Individual			Ambience		
			Exposure index (I.exp)			Pollution index (I.pol)		
			n*	Min–Max	Average	n	Min-Max	Average
Industrial	Cutting	1	/			1	—	0.0
	Stitching	1	/			1	—	0.8
	Uppers preparation	4	3	0.9–19.8	**8.4**	1	—	0.5
	Founder	7	5	0.3–3.5	1.5	2	0.3–0.8	0.5
	Finishing	3	3	0.9–1.8	1.2	/		
	Serigraphy	1	1	—	**9.4**	/		
Semi-industrial	Uppers preparation	8	4	0.5–2.5	**1.6**	4	0.1–7.2	**2.5**
	Founder	12	10	0.2–5.6	1.8	2	0.2–8.8	**4.5**
	Finishing	3	/			3	0.3–4.3	1.7
Artisanal	Uppers preparation	5	2	0.1–1.6	0.9	3	0.9–5.1	**2.5**
	Founder	8	5	0.4–6.4	**2.7**	3	0.4–0.7	0.6
	Uppers preparation/ founder	2	/			2	0.5–0.6	0.6
		Total = 33				**Total = 22**		

*Number of samples.

Table 3. Employee exposure index by workstation.

http://dx.doi.org/10.5772/intechopen.78316

Process	Workplace	n*	Acetone (mg/m³)		Cyclohexane (mg/m³)		n-Hexane (mg/m³)		MEK (mg/m³)		Toluene (mg/m³)	
			Min–Max	Average	Min–Max	Average	Min–Max	Average	Min–Max	Average	Min–Max	Average
Industrial	Cutting	1	—	0.0	—	0.0	—	0.0	—	0.0	—	0.0
	Stitching	1	—	0.0	—	160.5	—	36.8	—	0.0	—	6.8
	Uppers preparation	4	0.0–1652.0	413.0	0.0–183.8	66.8	28.1–975.0	**300.1**	0.0–2656.9	**776.6**	13.6–321.0	103.5
	Founder	7	0.0–3077.3	439.6	0.0–72.1	21.2	14.3–139.3	51.9	0.0–121.4	17.3	0.0–71.9	21.3
	Finishing	3	—	0.0	0.0–34.2	12.9	53.8–86.6	66.8	0.0–165.3	108.7	56–43.7	20.8
	Serigraphy	1	—	0,.0	—	**1281.1**	—	**401.8**	—	**1090.8**	—	38.6
Semi-industrial	Uppers preparation	8	0.0–114.3	16.1	—	118.5	0.0–204.0	57.4	0.0–509.5	113.4	4.4–500.0	106.6
	Founder	12	0.0–989.1	222.0	—	94.0	0.0–214.1	58.6	0.0–622.5	134.2	4.0–615.9	121.6
	Finishing	3	0.0–474.3	158.1	—	3.9	0.0–116.0	43.8	0.0–350.3	136.2	10.3–329.0	130.4
Artisanal	Uppers preparation	5	0.0–217.6	43.5	—	111.5	0.0–112.8	52.9	0.0–334.5	124.0	7.8–480.0	138.5
	Founder	8	—	0.0	—	112.2	3.3–147.3	58.6	0.0–442.5	150.8	0.4–578.1	115.0
	Uppers preparation / founder	2	00–43.4	21.7	—	26.6	21.6–27.2	24.4	—	0.0	27.7–35.9	31.8
		Total = 55										

*Number of samples.

Table 4. Average atmospheric concentrations of the solvents dosed.

and screen printing stations in the industrial process, the concentrations of the other solvents measured (acetone, cyclohexane, and toluene) were relatively high without exceeding the AEVs.

4. Discussion

This study is the first active sampling approach in the field of occupational health in Tunisia. It initiates the implementation of a structured approach in occupational toxicology and the environment by our laboratory in the theme of the impact of hazardous substances on the environment and human health. It required the acquisition of sampling equipment, which is of great interest to develop this type of toxicological measures and to carry out new measurement campaigns in Sfax city.

We had to experience a delay in carrying out our atmospheric sampling due to the various difficulties related to the development of the solvent analysis protocols. Indeed, we have carried out bibliographic research and repeated laboratory tests in the absence of technology transfer (North–South) methods of analysis [16, 29, 30].

For about 10 years the use of solvents is in full revolution, because of the constraints of PRP, but mainly because of regulatory requirements of the protection of the environment. These regulatory changes lead to changes in the nature of the solvents applied and in the way, they are used [11, 23, 31]. In addition, the number of employees exposed continues to increase in Tunisia. In France, the summer 2003 survey showed that the number of employees exposed to solvents has increased since 1994, from 12.2 to 14.7%, mainly in the chemical industry [2].

If these atmospheric metrology and toxicological analyzes were made for the first time in the footwear manufacturing sector in Sfax, they interested a sample of companies from the three footwear manufacturing processes (industrial, semi-industrial and artisanal) and were preceded by a preliminary RA with a post-study and an inventory of the products handled [5, 6]. This RE could be improved by the experience we have gained and with the best knowledge of this sector and its risks.

All samples taken from companies indicate that employee exposure to organic solvents varies widely depending on the task performed. Exposure indices were greater than 1 with VME overruns, particularly at the most exhibiting positions: the tiller, the smelter, and the finish. In our study, hexane levels were particularly high with averages ranging from 24.4 to 300.1 mg/m^3; while toluene levels were relatively high in some measurements but averages were below the VME range of 21.3–138.5 mg/m^3. In Spain and according to Cardona [32, 33], in his studies on the manufacture of footwear in the process the welded, by means of passive sampling (by badge) the average atmospheric concentration of hexane and toluene at the positions of founders, and the finish was, respectively, 47 and 86 mg/m^3 with a range of 4–652 and 2–1143 mg/m^3, respectively.

5. Conclusion

This study allowed us to know the chronic exposure to solvents in Tunisian shoes industry and to establish the first report on solvent exposure profiles in this sector. This exposure is not constant over time and varies according to the task performed and the manufacturing process. Therefore, the study of exposure to the workplace requires not only the average exposure to the AEV but also to identify the polluting phases to determine the short-term exposure to the AEV.

Following the phase of identification of solvents and the demonstration of exceedances of the limit values of certain solvents, the presence of women has increased in the shoe manufacturing companies, which encourages us to explore the reprotoxic nature of certain preparations. An approach to assess occupational risk through exposure bio-toxicological monitoring has thus been implemented in Sfax city in various sectors exposing solvents for a better POR.

Acknowledgements

The authors thank the University of Sfax for the financial support.

Conflict of interest

The authors declare that they have no conflict of interest in relation to this article.

Author details

Imed Gargouri[1,2]* and Moncef Khadhraoui[1,3]

*Address all correspondence to: imed.gargouri@fmsf.rnu.tn

1 Laboratory of Environmental Engineering and EcoTechnology (LR16ES19), Sfax University, Tunisia

2 Medical School, Sfax University, Tunisia

3 Higher Institute of Biotechnology of Sfax, Sfax University, Tunisia

References

[1] Dubeau M. Vapors and gas. In: Occupational Hygiene. 4th ed. Quebec, Canada: The Clay Griffin Inc.; 1985. pp. 415-523. ISBN 2-920210-34-3

[2] Arnaudo B, Leonard M, Sandret N, Cavet M, Coutrot T, Rivalin Thiérus L. Occupational risks in 2010: Large differences in exposure by sector. References Health Work. 2013. TF 207:59-74

[3] Wilson MP, Katharine Hammond S, Nicas M, Hubbard AE. Worker exposure to volatile organic compounds in the vehicle repair industry. Journal of Occupational and Environmental Hygiene. 2007;**4**:301-310

[4] Vincent R, Bonthoux F, Mallet G, Iparraguirre J-F, Rio S. Simplified chemical risk assessment methodology. Cahier note doc. 2005;**200**(ND 2233):39-62

[5] Gonzalez M, Velten M, Cantineau A. Evaluation of solvent exposure in an epidemiological study of 249 employees. Archives des Maladies Professionnelles et de l'Environnement. 1999;**60**:432-434

[6] INERIS (National Institute of Industrial Environment and Risks). Health Risk Assessment Guide. INERIS; 2003. 152 p [Online]. Available from: http://chimie.ineris.fr/en/lespdf/guide_ERS.pdf [Accessed: September 17, 2008]

[7] JORT. Law n ° 94-28 of 21 February 1994 on the compensation scheme for injuries at work and occupational diseases. JORT. 22 February 1994;**15**:308-18

[8] JORT. Law No. 95-56 of 28 June 1995 on the compensation scheme for injuries at work and occupational diseases in the public sector. JORT. July 4, 1995;**53**:1419-24

[9] Technical Center Leather Shoe Leather (CTC). Secrets of Shoe Making in Comics [Online]. Available from: www.ctc.fr/faq/questions.php3?theme=1 [Accessed: February 29, 2008)

[10] National Institute for Research and Security (INRS). Ventilation workshops gluing small objects (shoes). Practical Guide of Ventilation (n ° 5). 1987. ED 672: 28 p

[11] Poirot P, Hubert-Pelle G. Solvent exposure profiles and comparison with short-term limit values. ND 2235-200-05. HST (INRS). 2005;**200**:83-93

[12] French Society of Occupational Medicine. Medical surveillance of employees exposed to toxic substances for fetal development. Archives des Maladies Professionnelles et de l'Environnement. 2005;**66**:165-176

[13] Hertsenberg S, Brouwer D, Lurvink M, Rubingh C, Rijnders E, Tielemans E. Quantitative self-assessment of exposure to solvents among shoe repair men. The Annals of Occupational Hygiene. 2007;**51**:45-51

[14] Cocco P, Giuseppina Tocco M, Ibba A, Scano L, Grazia Ennas M, Flore C, Sanna Randaccio F. Trans, trans-muconic acid excretion in relation to environmental exposure to benzene. International Archives of Occupational and Environmental Health. 2003;**76**:456-460

[15] Armstrong TW, Caldwell DJ, Verma DK. A proposed methodology for determining exposure limits for hydrocarbon solvents. Occupy Day About Hygiene. 2005;**2**:600-607

[16] Mckee RH, Medeiros AM, Daughtrey WC. A proposed methodology for determining exposure limits for hydrocarbon solvents. Occupy Day About Hygiene. 2005;**2**:524-542

[17] Hervé-Bazin B. Guide for the evaluation of exposure to toxic risk in the workplace by sampling the atmosphere. ND 1730-135-89. Cahiers de Notes Documentaires. 1989;**135**:265-288

[18] INRS. Metrol Database [Online]. Available from: www.inrs.f [Accessed: March 20, 2006]

[19] IRSST (Robert Sauvé Research Institute for Occupational Health and Safety)–Operations Directorate. Sampling Guide for Air Contaminants in the Workplace. Canada: IRSST (8th ed.). T-06: 191 p. [Online] 2005. Available from: www.irsst.qc.ca [Accessed: August 15, 2006]

[20] IRSST. Substances sheet of the air contaminants sampling guide in the workplace. [Online]. Available from: www.irsst.qc.ca/en/_listersst.html [Accessed on 20.03.2006]

[21] SKC Inc. Air sample pumps, calibrators & accessories. Sorbent sample tubes and media collection [Online]. Available from: http://www.skcinc.com/ [Accessed: March 15, 2006]

[22] JORF. Decree No. 2007-1539 of 26 October 2007 setting binding occupational exposure limit values for certain chemical agents and amending the labor code (Part 2: Decrees in Council of State). JORF, October 28, 2007: Text 13 of 52 (NOR: TST0760893D)

[23] PLAY. Directive 2006/15/EC of 7 February 2006 establishing a second list of indicative occupational exposure lists under Council Directive 98/24/EC and amending directives 91/322/EC and 2000/39/EC. OF 9.2.2006: L36-L39 (EN)

[24] Pillary F, Conso F. Biotox: A Biotoxicological Guide for Occupational Physicians. France: INRS; September 2007. ED 791: 252 p [Online]. Available from: www.inrs.fr [Accessed: September 15, 2007]

[25] Triolet J. Occupational exposure limit values for hazardous substances. Values of ACGIH (United States) and MAK commission (Germany). ND 2114-176-99. Cahiers de Notes Documentaires. 1999;**176**:59-90

[26] Truchon G. Biological Monitoring Guide: Collection and Interpretation of Results. 6th ed. Canada: IRSST; 2004. T-03: 92 p. [Online]. Available from: www.irsst.qc.ca [Accessed: May 19, 2005]

[27] Courtois B. Occupational Exposure Limits for Chemical Agents in France. France: National Institute of Research and Security [INRS]; December 2007; ED 984: 19 p

[28] Poirot P, Subra I, Baudin V, Hery M, Chouaniere D, Vincent R. Determination of the medium-term exposure of house painters. ND 2125-179-00. Notebook Notes. 2000;**179**:5-13

[29] Eller PM, Cassinelli ME. NIOSH Manual of Analytical Methods (NMAM). 4th ed. USA: CDC; 1994. [Online]. Available from: www.cdc.gov/niosh/nmam [Accessed: August 15, 2006]

[30] Soulage C, Perrin D, Berenguer P, Pequignot JM. Sub-chronic exposure to toluene at 40 ppm alters the monoamine biosynthesis rate in discrete brain areas. Toxicology. 2004;**196**:21-30

[31] JORF. Decree No. 2008-602 of 25 June 2008 on the recycling and treatment of waste products. JORF. No. 149 of 27 June 2008 (NOR: DEVP0772293D)

[32] Cardona A, Marhuenda D, Marti J, Brugnone F, Roel J, Perbellini L. Biological monitoring of occupational exposure to n-hexane by measurements of urinary 2,5-hexadione. International Archives of Occupational and Environmental Health. 1993;**65**:71-74

[33] Cardona A, Marhuenda D, Prieto MJ, Marti J, Periago J-F, Sanchez J-M. Behaviour of urinary 2,5-hexanedione in occupational co-exposure to n-hexane and acetone. Int Arch Occup Environ Health. 1996;**68**:88-93

Chapter 2

Thermal Volumetric Analysis (TVA): A New Test Method of the Kinetics of Gas Emissions from Moulding Sands and Protective Coatings Heated by Liquid Alloy

Jerzy Stanisław Zych and Jan Mocek

Additional information is available at the end of the chapter

http://dx.doi.org/10.5772/intechopen.78369

Abstract

Production of medium-sized and large castings is carried out in the sand moulds. The sand moulds, which components are made of moulding sand, the binder are different kinds of materials, the most common: moulding resins (furan, alkyd, phenolic, etc.). Internal surfaces of moulds, for increased thermal resistance, are usually covered by protective coatings. Liquid foundry alloys filling sand moulds produce intense heat, especially in layers adjacent to the cast. Under such conditions, the organic binders are burned or gasified. The resulting gases can penetrate the liquid causing the formation of defects such as blow-holes, pinholes, pitted skin, etc. The greater the intensity of the secretion of gases and the greater their volume (number), the more likely defects are created in castings. An important quality criterion of foundry resins is their tendency to emit gas. The authors developed the new test method, which allows to monitor the kinetics of gas emissions from the moulding sand and foundry coatings as a function of time and temperature (thermal volumetric analysis (TVA)). A number of studies were made for moulding sand with organic binders. Description of the research methodology and the presentation of results are the subject of the article.

Keywords: thermal volumetric analysis, moulding resins, moulding sands, protective coatings

1. Introduction

Heating of foundry moulds by liquid metal filling the inlet system and mould cavity causes often heating of the successive - deeper and deeper situated - layers of the mould, to a very high temperature. In case of cast steel, a temperature of layers adjacent to the casting surface often exceeds 1000°C. In case of other alloys, heating is also very intensive; however, such a high temperature is not reached. A part of components of moulding and core sands and protective coatings, applied in the casting technology, undergoes gasification (burning or distillation). Gaseous products heated to high temperatures constitute an additional volume of gases, apart from the heated air. These gases should be removed from the mould cavity before it will be filled with metal. When they are not completely removed the gas pressure increase is observed in the mould cavity and in moulding (core) sand layers adjacent to the mould cavity. This often causes the penetration of gases into liquid and solidifying metal, which leads to casting surface defects classified as originated from gases. They are in a form of pinholes, blow-holes (internal or external), or shrinkage gaseous porosity. Examples of castings with surface defects are shown in **Figures 1–5**.

A trend of formation of this group of defects depends on several factors, but nearly always grows with an increasing gasification of moulding sands and protective coatings. The gasification is understood as a tendency of the material to form gases when it is heated. Therefore, investigations of the moulding sands gasification, understood as the amount of gases emitted by the mass unit of the material heated to app. 1000°C are often performed. The total amount of gases emitted from the moulding sand is important and has an essential influence on forming the described defects. However, the instantaneous values of gas pressures-at the metal and mould boundary-decide whether gaseous defects (shown in **Figures 1–5**) will appear. The higher is the gas pressure in the mould, during the initial phase of alloy solidification,

Figure 1. Iron casting with visible blow-holes on the surface, produced in the mould of wet bentonite moulding sand high-pressure technology.

Figure 2. Pump body casting of Al-Si alloy with visible blow-holes on the surface, produced in the sand mould - hand moulding technology.

Figure 3. Casting of spheroidal cast iron with the visible defect of the gaseous origin, called pitted skin, produced in the wet bentonite mould - machine moulding technology [1].

Figure 4. High-dimensional iron casting with visible gaseous defects on the surface, produced in the sand mould-hand moulding technology with the application of moulding sand with furan resin.

Figure 5. Casting of Al-Si alloy with visible blow-holes on the surface, produced in the metal mould-low-pressure casting technology.

the more probable is the defect formation. Thus, in assessing the gasification of materials applied in the moulding sand technology, there is a necessity of determining the gas emission kinetics in relation to the heating time and-even better-as the function of the instantaneous temperature at which they evolve.

The results of the gas emission investigations realised according to the new concept and with the application of the new measuring method, which allows for the parallel measuring of gas volumes in the time unit in relation to the instantaneous temperature at which the process is performed, are presented in this chapter.

1.1. Analysis of the problem and the research stand construction

Evolutions of gases from heated materials are physical processes related to the decomposition of components, their distillation or burning. These processes are accompanied by the sample mass loss and thermal effect (emission or absorption of thermal energy), leading to the instantaneous temperature increase or decrease. For investigating the described processes the thermogravimetry (TG) is applied, while when the mass change rates of heated samples are tested the derivative thermogravimetry (DTG) is used. The TG curve constitutes the function expressing changes of the sample mass (Δm) in dependence of time (t) or temperature (T) [2]:

$$\Delta m = f(t) \rightarrow f(T) \rightarrow TG \tag{1}$$

The first derivative of this function versus time, expresses the mass change rate:

$$\frac{d(m)}{d(t)} = f'(t) \rightarrow f'(T) \rightarrow DTG \tag{2}$$

The DTG curve is characterised by the maximum and minimum peaks, occurring at the determined temperature values. In these points, the process rates are the highest. Devices called derivatographs are simultaneously used for recording the TG curves and thermal analysis (DTA) of the tested material. The schematic presentation of such device is shown in **Figure 6**.

The analysis of the mass changes kinetics of heated materials, applied in the mould casting technology, is not sufficient for assessing their technological suitability. The analysis of the

http://dx.doi.org/10.5772/intechopen.78369

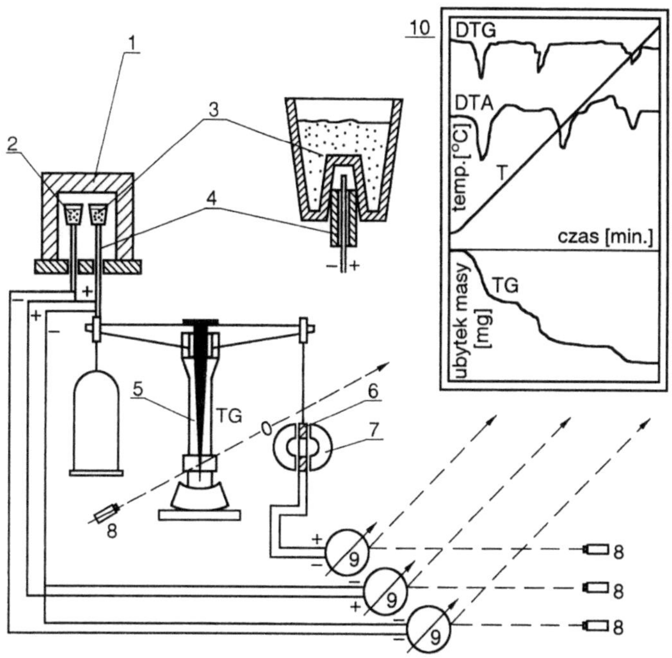

Figure 6. Schematic presentation of the derivatograph: 1 - furnace, 2 - crucible with the standard material, 3 - crucible with the tested material, 4 – Thermocouples, 5 – Scale, 6 – Induction coil, 7 – Magnet, 8 – Lamp, 9 – Galvanometer, 10 – Thermobalance [2].

amount and kinetics of evolved gases, related to the sample mass change, is also necessary. No simple quantitative relationship exists between the kinetics of the evolution of gases and the kinetics of mass changes of heated materials undergoing a decomposition (burning). It is not possible to calculate the volume of gases evolved during heating, from the mass changes of the material, which was thermally decomposed.

The authors of the hereby paper developed the investigation method of the thermal decomposition of materials in which measurements of instantaneous amounts of gases evolved from the sample in the heating process, instead of the instantaneous sample mass changes, are performed. The temperature of the tested material sample is measured simultaneously. In analogy to the TG this new investigation method was named thermal volumetric analysis (TVA) – investigations of the kinetics of the evolution of gases from materials subjected to the thermal decomposition. In analogy to Eqs. (1) and (2), descriptions of the volume changes of gases generated in the thermal decomposition process of materials are of the general form of Eqs. (3) and (4).

$$\Delta V = f(t) \rightarrow f(T) \rightarrow TV \tag{3}$$

$$\frac{d(V)}{d(t)} = f'(t) \rightarrow f'(T) \rightarrow DTV \tag{4}$$

where ΔV – volume of gases, formed as a result of the thermal decomposition (burning) of the material.

Materials for casting moulds and cores are assessed in respect of their gas forming ability. The most often, in up to now investigations, the measurements of the total volume of gases evolved from the unit mass sample [2–7] were performed, while a part of them was performed with the determination of the evolution rates in the successive time heating intervals [2, 8–11]. Thus, only dependencies that constitute the first members of Eqs. 3 and 4 were determined. It was only when the investigations described in the chapter of authors [12–16] were published that it was possible to assess the amount and rate of gases evolution in relation to a temperature. These measurements can be considered as the beginning of the derivative TVA - investigations of the kinetics of the evolution of gases from casting materials as the heating temperature function (TV). In a similar fashion as in the TG, where the most important structure element is the moulding sand probe (crucible) equipped with thermocouple (item 3 in **Figure 6**), in the thermal volumetric method the most important element of the research stand is the test probe equipped with the temperature sensor (B2) shown in **Figure 7**.

The test probe (B) in which the sample of the tested moulding sand is placed, was designed and built as unilaterally closed measuring flask of a pipe shape. To increase its resistance to thermal shocks the test probe was made of heat-resistant steel. On the other side, the test probe was equipped with a sealed valve (**Figure 7B4**). This closing valve allows for directing the gas, evolving from the sample to the measuring system. The same valve, after being disconnected from the probe (pipe), allows for the analysed moulding sand replacement (P). After performing the measurement the test probe is emptied of the tested (heated) moulding sand and purified from its remains. To reduce the test probe inner space filled with air, an insert is placed inside it (**Figure 7B3**). On the outer surface of the test probe, the pipe (B2) - guide bar - with ceramic insulators inside was placed. In these insulators, the thermocouple conduits of the 'S' type (Pt–PtRh10) (B1) were interlaced. A measuring junction of the thermocouple is placed in the zone where is the tested moulding sand sample (P). Such solution ensures an accurate and continuous temperature control of this moulding sand (P). The thermocouple conduits being at the end of the test probe were connected to the compensation conduit (B6) and to the temperature recorder.

The schematic presentation of the new stand for testing the gas evolution kinetics and for assessing gas generations of moulding sands-within TVA is provided in **Figure 8**. The test probe with the tested moulding sand (B) is introduced into the tubular style furnace PSR-1 (**Figure 8A2**) with a temperature controlled by means of a thermocouple (A3) as well as the

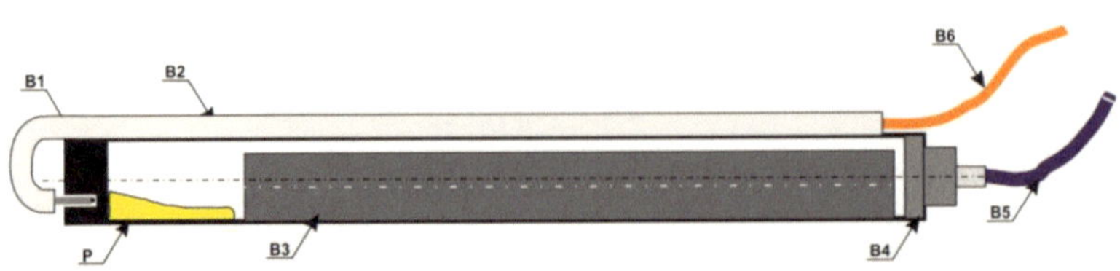

Figure 7. Structure of the measuring sensor (B) for the thermal volumetric investigations [12].

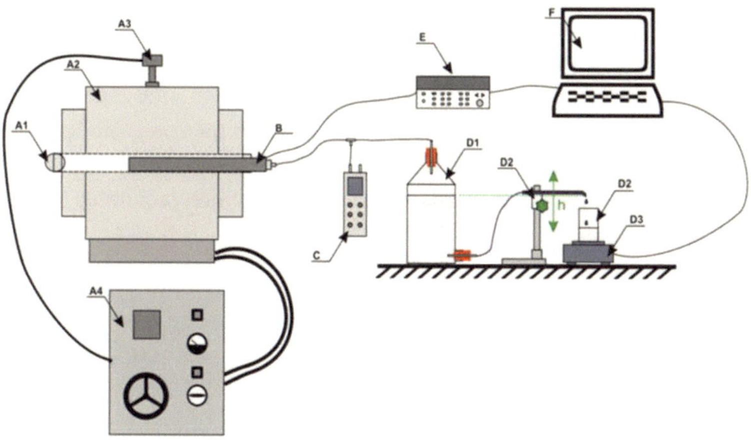

Figure 8. Schematic presentation of the laboratory stand for testing the gases evolutions kinetics and for assessing gas generations of moulding and core sands [12].

power supply and digital regulator (A4) systems. The back side of the furnace operating pipe is closed by thermo-insulating wool (A1) to limit temperature drops caused by the air flow. The air filling the cold test probe (increasing its volume as a result of heating), as well as the gas, originated from the decomposition of the tested moulding sand binder are directed-via a complex of conductors (B5)-to the system measuring their volume. The system comprises a three-way pipe distributing gas to the low-pressure meter (**Figure 9C**) and to the glass container with water (D1).

In the bottom part of the vessel with water, there is a drain, closed with a plug with a drain pipe. Gas originated from the thermal decomposition of the tested material (from the test probe **Figure 7**) is introduced via the top valve. This gas flowing into vessel D1 pushes out water from it in such volume as the gas itself occupies in the vessel. Water flowing out via the bottom valve of the vessel is first directed, by a rubber hose, to the pressure 'controller' (hydrostatic pressure) of the stand (D2) and then to the outlet zone of the pipe. It flows out from the pipe in a form of individual drops with the intensity corresponding to the gas evolution rate.

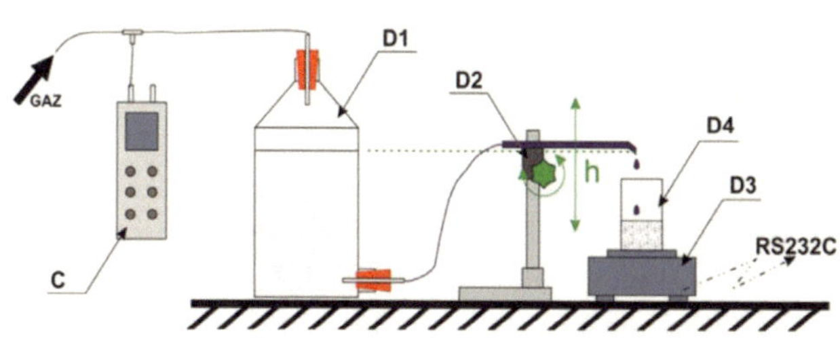

Figure 9. Schematic presentation of the measuring section of the gas evolving from the heated sample of the tested material [12].

Under the outlet zone of the pipe, the vessel (D4) was placed on the tensometric laboratory scale, equipped with the electronic system for the data collection and transmission to the computer. The thermocouple placed in the test probe (**Figure 7B**) is connected to the computer via the recorder Agilent 34970A (**Figure 8E**). During the measurements, the data concerning the temperature and volume of gases evolving at a particular time are gathered. Due to the fact that the water specific density equals 1 g/cm^3, the number of grams of water pushed out by the gas flowing into the vessel corresponds to its volume, which facilitates the measurement-calculation procedure.

Measurement method: The test probe filled with the tested material sample (e.g., moulding sand) is introduced into the furnace heated to a predefined temperature. The sample is weighed by means of the laboratory scale, with an accuracy of up to 0.01 g. The sample quantity is selected in a way assuring uniform heating in the whole volume, namely, it cannot be too large. On the other side, it should not be too small because small amounts of evolved gases make difficult their accurate measurements. At the determined measuring accuracy of devices (tensammetric scale), measuring of small values (water pushed out from the vessel) are burdened with a higher relative error.

A thin layer of the moulding sand sample assures fast heating of its total volume and simultaneous gasification (burning) in the entire volume of the tested material. The weighted portion of the moulding sand is put into a clean test probe, and then a roller (B3) is inserted to reduce a 'dead volume' of the measuring system (test probe). After connecting the pipes, the water level in the vessel and outflow are equalised-the location of a drain pipe outlet. After water levelling the manometer – shows 0 mm of the water column. When the pressures are equalised the software recording the volume of water pushed out from the vessel and recording the temperature of the heated moulding sand sample is activated. The gas source constitutes the tested sample undergoing a partial gasification. When the test probe is heated the gas volume in it increases, which while flowing out to the vessel with water induces the commencement of water flowing out. At the same time in the 'on-line' mode, the quantity of outflowing water as well as the temperature of the heated moulding sand is recorded.

The gathered data are then subjected to treatment and the results are the most often presented in forms of diagrams.

Test probe calibration: the test probe volume in the described version equals app. 40 cm^3. When the test probe with the tested moulding sand is heated two processes simultaneously occur: increasing - due to heating - of the air volume filling the probe and the gas evolution from the tested material (e.g., moulding sand), which is subjected to distillation, burning, etc. The measurement of the total gas volume pushed out from the container provides the erroneous gasification value of the tested material. It could be called either *the total gasification or apparent gasification*. The total volume of gases flowing out from the container depends on the proportion between the working volume of the test probe and the moulding sand sample volume (weight). The higher working volume of the test probe the more 'erroneous' is the picture of the real gasification of the moulding sand.

In order to make possible the determination of the real amount of gases formed due to the moulding sand decomposition, it is necessary to determine the volume increase of the air heated to the selected temperature in the test probe. This measurement can be called the test probe 'calibration' since it is based on recording the increasing gas (air) volume during heating the test probe to the selected temperature. To perform the measurement of the real gas evolution from the tested moulding sand sample the calibration results should be subtracted from the recorded amounts of gases evolving from the sample.

The obtained results should be divided by the amount of the tested moulding sand. Then the achieved result will be the quantity of gases emitted by 1 g of the moulding sand at the given temperature. In case of materials consisted of the matrix (which is not decomposed) and the binder, it is possible to recalculate the results into the weight fraction of the binder itself. For this purpose, the moulding sand composition and the binder percentage fraction should be known. In such case it is possible to refer to the specific gas evolution, it means to 1 g of the binder contained in the moulding sand (not to 1 g of the moulding sand).

2. Thermal volumetric analysis applied for testing the materials used in the casting moulds technology

2.1. Kinetics of gases evolutions from core sands

A lot of castings produced in sand or metal moulds require applications of cores for the reproduction of internal shapes of castings. Due to the fact that casting cores are poured with metal nearly from all sides, their tendency for the evolution of gases is very important since very often these gases are retained in casting walls, forming defects. Especially in such cases, a low gas generation tendency of core sands is very important. The example of such casting can be the element of a bathroom battery. The half of the metal mould with the inserted sand core is shown in **Figure 10**. In this mould-after joining its two halves-the casting of the element of the domestic plumbing fittings, the so-called spout, will be made of brass. The sand core is produced in the hot box technology in which the core sand binders are thermohardened resins, the most often phenol or phenol-urea resins.

Investigations applying the new method thermal volumetry analysis were performed to assess the quality of moulding sands used for cores, from the point of view of their tendency for the evolution of gases during heating by liquid metal. The results of these tests are presented in **Figures 11–14**. Cores were produced of sands in which thermohardened resin, classified to the phenol group, was used. The core sand contained approximately 2.0% of phenol-urea resin, Resistal HB AT 12. Tests of gas evolution rates were carried out on the research stand, which scheme is shown in **Figures 8** and **9**. As it was already indicated, during investigations of the moulding sand thermal decomposition and the evolution of gases accompanying this process-performed by the new method-the continuous measurements of the sample temperature and the volume of gases generated by the heated moulding sand were carried out. In addition, the test probe calibration described above was performed. The obtained results are presented in **Figure 11**. On their bases, according to general Eq. 4, the gases evolution rates as the heating time function (**Figure 12**) and as the material (in this case the core sand) sample TV (**Figure 13**) are determined. Kinetics of the gases evolution shown in **Figure 12** indicates that this process is intensive already in the first minute of heating. The sand without additional drying emits much more gases and much earlier, which favours the gaseous defects formation.

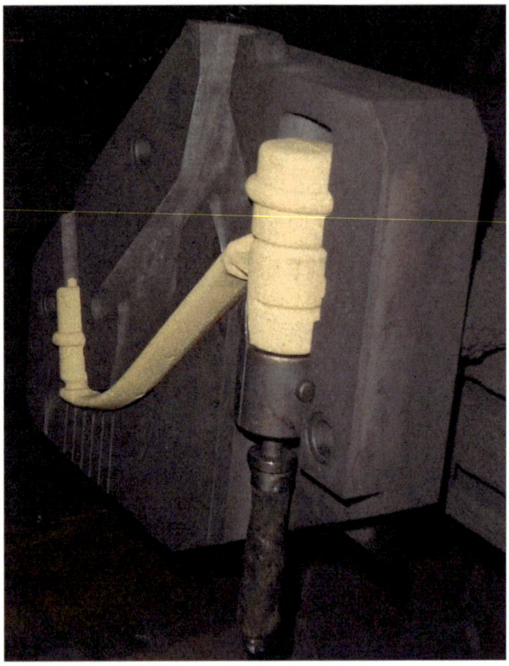

Figure 10. Metal mould (half of the mould) with the sand core.

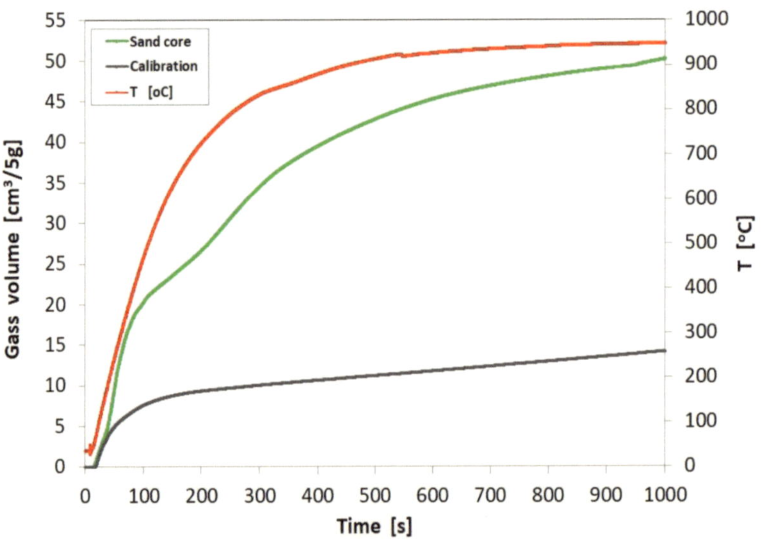

Figure 11. Changes of the sample temperature and of the volume of gases, recorded during the test probe calibration and core sand measurements (without drying) for the hot box technology.

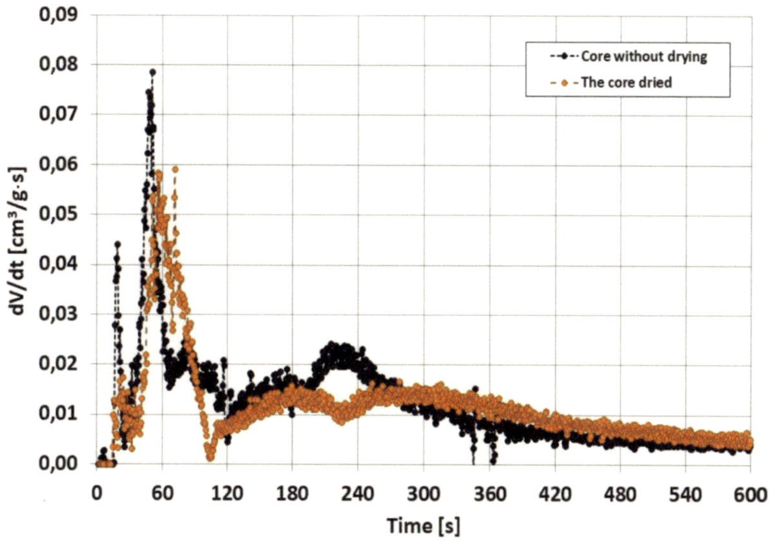

Figure 12. Kinetics of the evolution of gases from the core sand heated to app. 950°C.

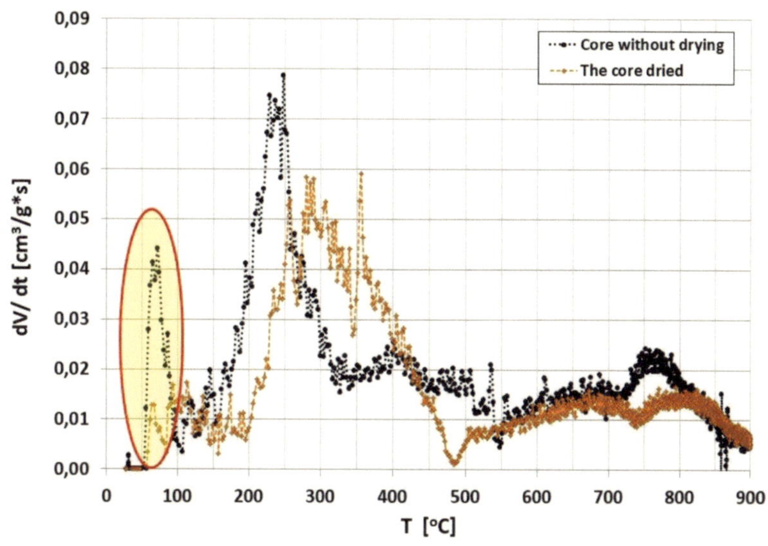

Figure 13. Intensity of the evolutions of gases from the heated core sand, applied in the hot box technology, as the heating TV.

Courses of the evolution of gases from cores without the previous drying and after additional drying are compared in **Figure 13**. The core was dried for 1 h at a temperature of 100°C. Three maxima can be separated in the courses of the evolution rates ((dV/dt) = f (T)): the first one, when a temperature is near 100°C; the second, when the sample temperature is within 250–300°C range and the third, when the temperature is nearing approximately 800°C. Gases from the core can penetrate the casting only during the period when metal is

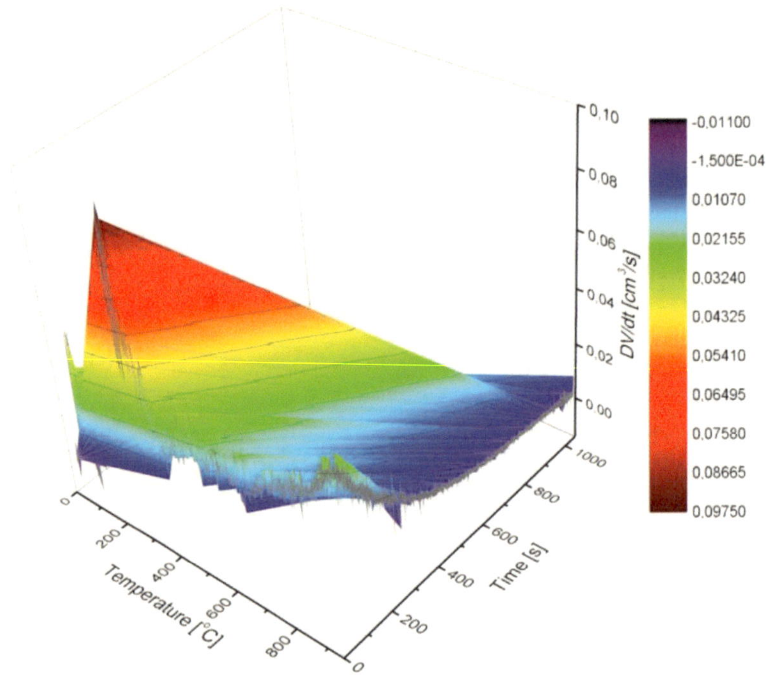

Figure 14. Intensity of gases evolutions from the core sand as the TV and the heating time function.

in the liquid state. In case of the casting shown in **Figure 10**, this period is no longer than app. 10 s. Therefore, for the production technology of these castings, two first peaks-seen in **Figure 13**-are important. Especially dangerous is the occurrence of the first maximum since in this period metal poured into the mould is still in the liquid state. The practical conclusion from the described example of the TVA is such: the cores produced in the hot box technology with using phenol-urea resin as a binder should be dried, to get rid of the first peak of the volume of gases seen in **Figures 12** and **13**.

The developed investigation method allows to analyse the process of evolution of gases from the selected material either as the heating time function (**Figure 12**) or as the TV (**Figure 13**). These two parameters: temperature and time cannot be physically separated since the temperature changes occur in parallel with the heating time changes. However, on the bases of the same data, the spatial picture of the gases evolution intensity from the heated material can be built in the real-time function and the real TV. The example of such approach to the description of the gases evolution from the tested core sand, for which partial dependencies are shown in **Figures 12** and **13**, is presented in **Figure 14**. It can be noticed, that the intensive evolution occurs at the very beginning of the heating process and at a relatively low-temperature. At further heating, regardless of the higher temperature, the intensity of the gases evolution decreases. When the tested material was approaching this higher temperature a large amount of the binder (resin) was already gasified. This particular conclusion concerns only the tested core sand with the phenol-urea resin. Other materials will be forming a different picture of the gas evolution courses.

2.2. Kinetics of the gases emission from protective coatings

Inner surfaces of sand moulds and cores, in order to increase their thermal resistance, are often covered by various protective coatings. A potential source of gases in protective coatings is diluents (alcohols, water), which, regardless of drying of coatings deposited on moulds, are not always completely removed. Water in coatings can occur as a diluent (free water) or as a component of material used for making the coating (e.g., bentonite) and in such case, it is the so-called bound water. To improve rheological properties of coatings various additions are often introduced (e.g., polymers), which can also be the source of gases originated during the mould heating after covering it with liquid metal. Also, metal moulds are covered by protective coatings to decrease the casting cooling rates, for the easier separation of castings from moulds and to protect the mould surface against thermal shocks. However, at the increased inclination of coatings for the gases evolution, casting surface defects can be formed (**Figure 5**).

In the previous solution, the assessment of the coatings inclination to evolve gases during heating was performed on the bases of DTG investigations. Information concerning the temperature at which the heated material sample decreases its mass, due to the gasification of some of its components, was obtained from these tests. The scale of the sample mass changes could be utilised for predicting amounts of gases, by means of proportion: larger changes of the sample mass-larger volumes of gases. During the DTG tests, the heating process of the sample is performed with a constant and usually small rate. Heating of the casting moulds elements is abrupt. The research stand for the thermal volumetric tests described in the hereby paper is prepared for very fast heating of the material sample, which corresponds to the conditions occurring in casting moulds after being poured with liquid metal. To compare both methods assessing the thermal decomposition of materials, one protective coating was tested by means of the TG, DTG and TV, DTV methods. Schematic presentations of the research stands for both methods are shown in **Figure 15**, indicating that the main difference constitutes the construction of test probes: one is used for testing changes of the sample mass

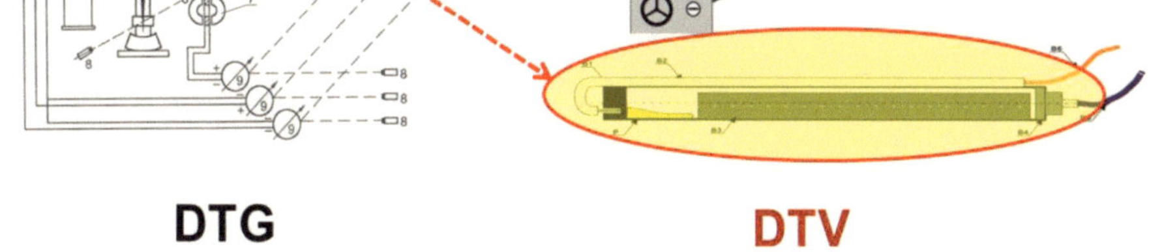

Figure 15. Schematic presentation of the research stands for the DTG and thermal volumetric analysis (DTV).

DTG, while the other one – for testing the volume of gases evolving from the heated material (DTV). If the heating rates were the same on both stands the diagrams of mass changes and of volume changes of gases, presented as the TV, would be very similar.

Since the situation is not like that, in order to find differences the comparable investigations of one protective coating were performed by both methods. Investigations concern the alcoholic coating, called Velvacoat GH70, applied for sand moulds and cores. The results of thermogravimetric tests (DTG) are seen in **Figure 16**, while the results of investigations of evolutions of gases and their rates (DTV) in **Figure 17**. In the qualitative approach, a

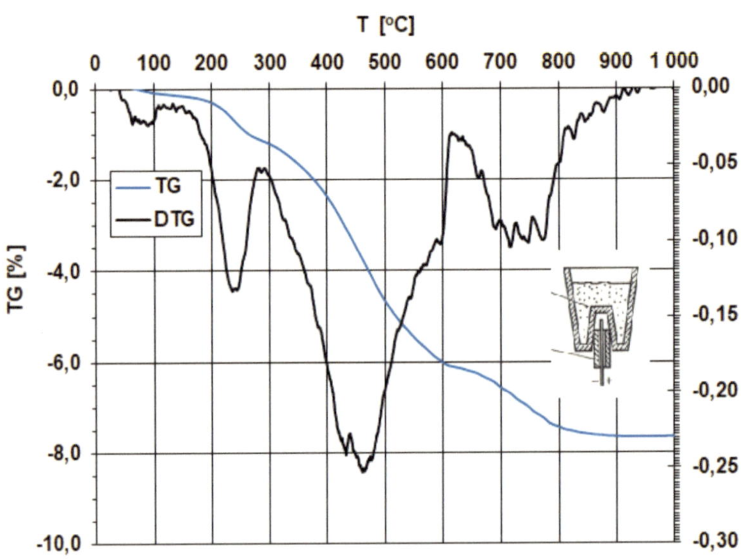

Figure 16. Investigation results (thermogravimetry DTG) of the alcoholic coating (Velvacoat GH70).

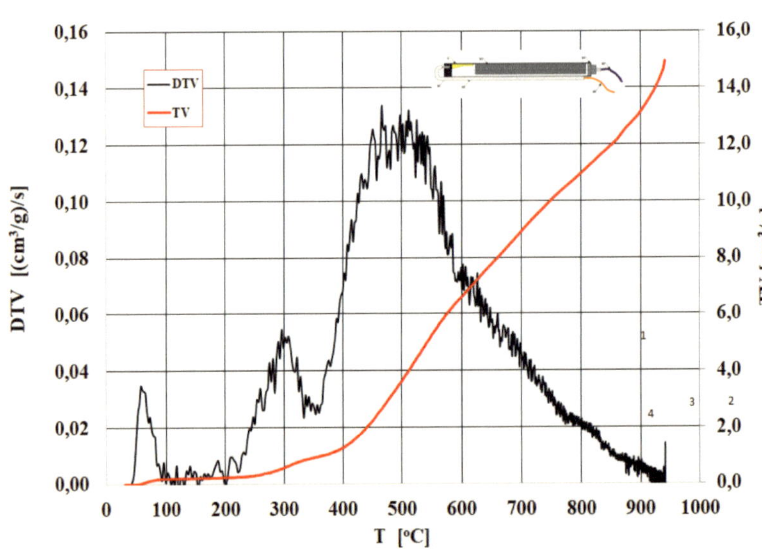

Figure 17. Investigation results (thermal volumetric analysis DTV) of the alcoholic coating (Velvacoat GH70).

significant similarity of the DTG and DTY curves can be noticed since in both cases peaks of changes of derivative values occur at similar temperature values. The first peaks occur at a temperature app. 100°C, the second at 250–300°C and the third, the highest, at temperatures 450–500°C. The fourth peak occurring in the TDG curve at temperatures 700–800°C is not clearly recorded in the TDV line. The results obtained from comparing these two methods indicate that the assessments of the thermal decomposition coincide with each other. However, the new investigation method, thermal volumetric analysis, provides additional important information on materials, essential for several technological production processes. The intensive evolution of gases from heated materials renders difficult the technologies of casting production regardless of being realised in sand, ceramic or metal moulds.

The thermal volumetric method was also applied in investigations of other protective coatings used for covering sand cores. During the preparation of coatings, they are usually diluted with alcohol, which vapourises after coatings are deposited on the core surface. To get rid of the diluent, before the gas evolution measurements, coatings were dried for 2 h at a temperature of 130°C. After drying the coating was crushed to powder. Four coatings of symbolic designations were tested. These coatings - after drying - are presented in **Figure 18**.

The gases evolution courses, during heating, are presented in **Figures 19** and **20**. The gas evolution courses as the TV are shown in **Figure 19**. It can be noticed, that peaks occur in two temperature ranges: 250–350°C and 550–600°C. Up to the present convictions, that only not properly dried coatings can be the source of gases in casting moulds after they were poured with liquid metal, are not confirmed. All tested coatings, regardless of being dried, emit significant amounts of gases (**Figure 19**).

From the point of view of the technological suitability and maintaining a low risk of casting defects forming, the coatings characterised by the lowest gas emission should be selected. The second recommended feature is shifting of peaks of the most intensive emissions to higher temperature values. In such case, it is more probable that metal will solidify earlier than large amounts of gases from the coating will appear in the mould. Out of the compared coatings the lowest inclination for evolving of gases has the coating marked with BC3 symbol. The result of these investigations was positively verified in one of the Polish foundry plants. The developed new method of investigating the materials inclination for emissions of gases when they are heated (thermal volumetric analysis DTV), becomes the good tool for assessing the technological suitability of several materials, including protective coatings.

Figure 18. Alcoholic coatings, after 2 h of drying at 130°C.

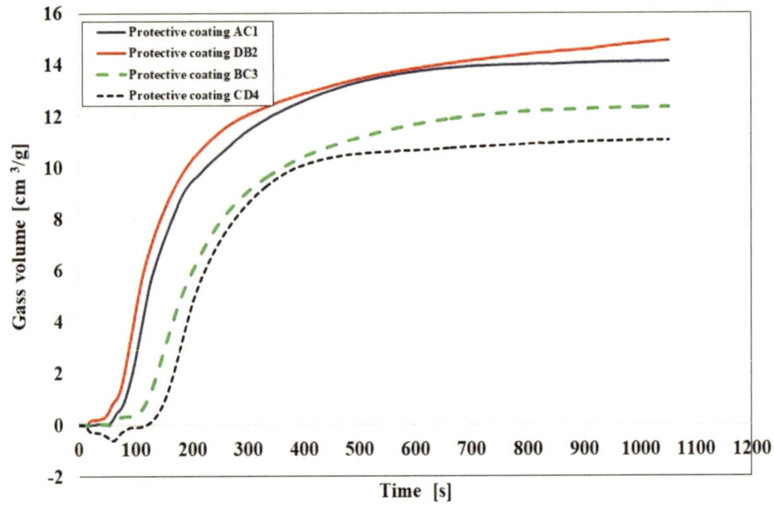

Figure 19. Gases evolution courses from alcoholic protective coatings applied for sand moulds and cores.

2.3. Kinetics of gases evolutions from expanded polystyrene

Currently the full mould technology in several versions is more often applied in the foundry practice. This technology essence comes down to using casting patterns made of expanded polystyrene. The pattern is expendable and after making the mould remains in it. The poured metal causes its gasification and by taking its place obtains the shape of the casting. The most important in this technology is the amount of gases originated from the casting pattern gasification since it decides on the possibility of producing castings without defects. Equally important parameter, possibly even the most important one, is the intensity of the evolution of gases during pouring. In one technology the polystyrene pattern is covered by the coating layer, and after drying the whole is placed in the box with the dry sand. Gases from the polystyrene gasification process have to permeate at first through the protective coating and then through the sand, to be finely outside the mould. Too slow flowing out of gases from the inside of the mould can be the reason of the liquid metal gasification, changes of its chemical composition and the porosity formation [7]. Within the project of assessing materials for patterns for the full mould technology the kinetics of gases evolutions from three kinds of gasified materials was investigated. Investigations comprised: polystyrene of a white colour ($\rho_o = 23.2$ kg/m^3), blue ($\rho_o = 39.4$ kg/m^3) and extruded polystyrene Austrotherm XPS TOP 50 SF ($\rho_o = 47.8$ kg/m^3) of a pink colour.

The results of gas evolutions of the tested polystyrenes are presented in **Figures 21–24**. The source material of the performed measurements is the record of temperature changes of the heated polystyrene sample and the record of the amount of gases emitted by the sample in successive seconds of its heating. Such record is presented in **Figure 21**. The amount of the evolved gas can be referred either to the sample mass (1 g) or to its volume (1 cm^3). Referring to the material volume is more justified in the foundry practice. In such approach, in relation to the volume unit, the smallest amounts of gases emits white polystyrene, slightly more blue

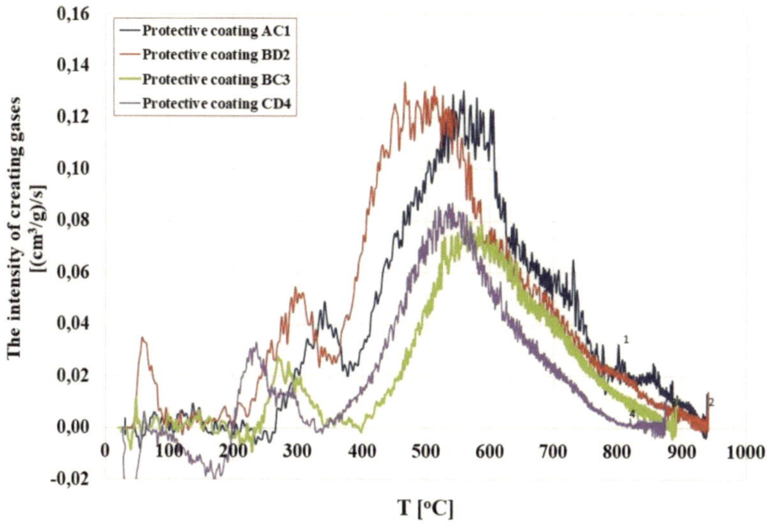

Figure 20. Intensity of gases evolution (DTV) from protective coatings applied for sand moulds and cores.

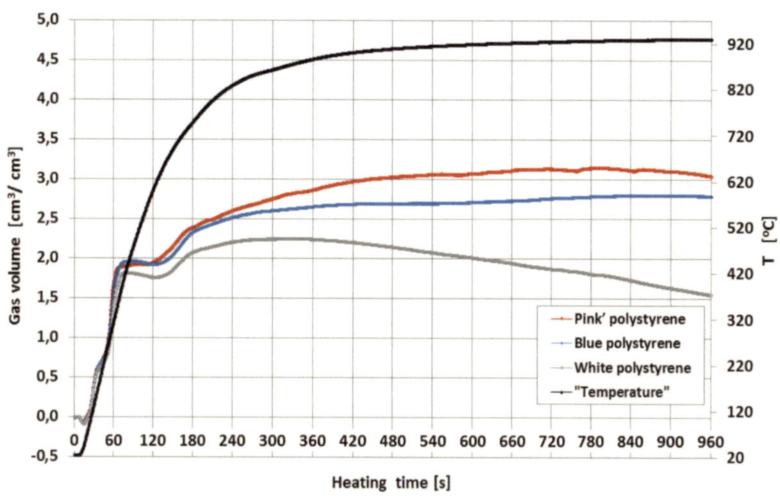

Figure 21. The course of heating of the expanded polystyrene samples and of the emission of gases.

polystyrene, while the highest amounts the 'pink' one. Differences reach more than 20% of the volume. Each of these polystyrenes has a different density, which is related to the amount of emitted gases. Apart from the amount the important factor is the rate of the gasification process of polystyrene. For 'white polystyrene' amounts of gases emitted in successive seconds of heating are shown in **Figure 22**. The remaining materials have similar courses of the emission of gases. Three, different sizes, peaks of the gas evolution intensity can be singled out in the gasification process. Two first peaks are larger and occur one after another in a short time (after 30 and 60s), and the third one, significantly smaller, occurs much later, after approximately 160 s.

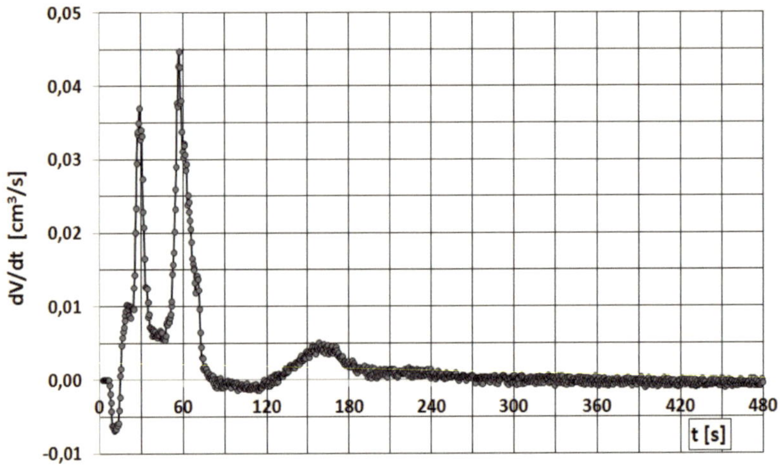

Figure 22. Kinetics of the emission of gases from expanded polystyrene (white) during its gasification (heating of 1 cm^3 of polystyrene).

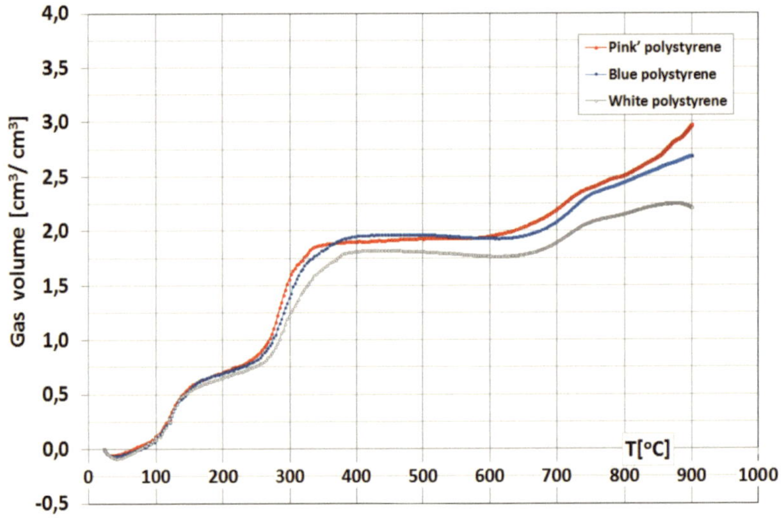

Figure 23. Gases evolutions during heating the polystyrene samples, the influence of a temperature on the amount of gases emitted from 1 cm^3 of polystyrene.

Assignments of temperatures, at which successive changes of the gases evolutions rates occur, are visible in **Figures 23** and **24**. Three temperature values, at which the acceleration of the gas emission process is quite visible, are seen in **Figure 23**. It can be seen more accurately in **Figure 24**. Applying the developed thermal volumetric method, derivatives dV/dt were determined and the temperature values were assigned to them (DTV). The first maximum of the gases emission rate occurs at a temperature of app. 130–140°C, the second - at a temperature of app. 300°C, while the third, significantly smaller, at a temperature from the range: 700–750°C.

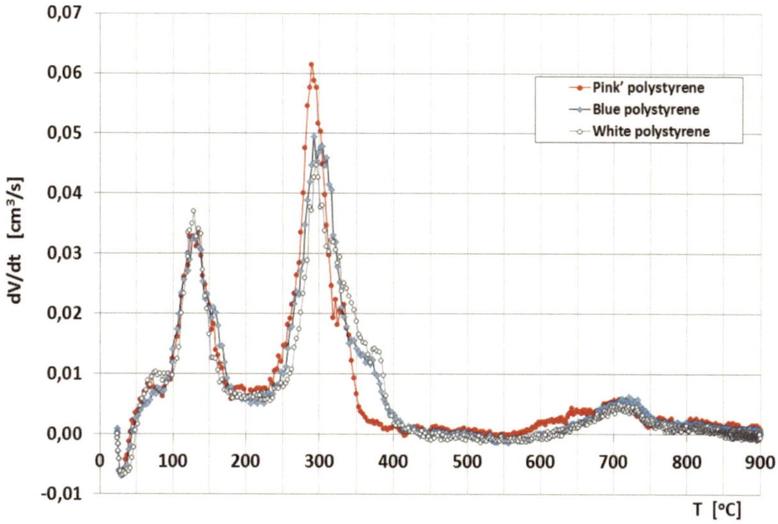

Figure 24. Influence of the temperature of polystyrene heating on the gases emission kinetics.

These are the temperature ranges and values at which the gas pressure in the mould rapidly increases and the mould (especially its coating) must be able to carry away these fast increasing amounts of gases. The gasification process of polystyrene is the most intensive at a temperature of app. 300°C. When comparing kinetics of gas evolutions from the expanded polystyrene (**Figure 24**) and from resin (**Figure 13**) their similarity can be noticed. This similarity is due to the fact that these both materials are organic compounds.

3. Conclusions

The presented new method of investigating the kinetics of gases evolutions from materials subjected to heating, called the thermal volumetric analysis (DTV), constitutes the supplement of the up to now applied thermo-gravimetric (DTG) method of testing the kinetics of the sample mass changes of the heated materials. As it was revealed in investigations both methods are compatible with each other, the thermo-gravimetry allows to determine temperature values at which the thermal decomposition and gasification of the material components occur, while the thermal volumetric analysis allows to determine the intensity of the emission of gases at the same temperature values. Tests carried out for several materials and plastics applied in casting technologies confirmed the high suitability of this method, and the obtained results can constitute the base for assessing the suitability of the given material for the casting process.

This new investigation method is currently at the stage of testing prototype solutions in construction of the research stand and is, within this zone, improved. The range of the future applications of the thermal volumetric analysis should not be narrowed to materials used in foundry practice only. The problem of the intensive emission of gases from heated materials occurs in several processes and often complicates various technologies.

Author details

Jerzy Stanisław Zych* and Jan Mocek

*Address all correspondence to: jzych@agh.edu.pl

AGH University of Science and Technology Faculty of Foundry Engineering, Kraków, Poland

References

[1] Bajer J, Köppen M. Manual of Casting Defects. 1st ed. IKO – ERBSLÖH Industrrieminer-alien and Kohlestoff GmbH -Marl; 1994

[2] Lewandowski JL. Tworzywa na formy odlewnicze. Kraków: Akapit-Krakow; 1997. ISBN: 83-7108-21-1

[3] Lewandowski JL, Solarski W, Pawłowski Z. Classification of moulding and core sands as regards gas formation. Przegląd Odlewnictwa. 1993;**5**:143-149

[4] Urbanik E. Study of the kinetics of gas extraction of dried organic adhesives moulding sand. Doctoral Dissertation, Kraków: AGH, Faculty of Foundry Engineering; 1964

[5] Żółkiewicz Z. Kinetyka cieplnego rozpadu modelu styropianowego oraz wpływ wydzielających się gazów na zawartość wodoru, azotu i tlenu w żeliwie szarym, w "procesie pełnej formy". Doctoral Dissertation, AGH, Faculty of Foundry Engineering, Kraków; 2001

[6] Pacyniak T. Metoda pełnej formy. Wybrane aspekty, (Method Lost Foam Casting. Selected Aspects), Copyright by Politechnika Łódzka; 2013. ISBN 978-83-7283-555-0

[7] Bates CH, Littleton H, Stroom P, Mcmellon B. Technological Developments and Worlwide Market Growth in Lost Foam Casting Production; 64th World FoundryCongress, Paris 2000. Referat UA

[8] Holtzer M, Dańko J, Lewandowski JL. Station for research of the volume and harmful-ness of gases compounds from the materials used in foundry and metallurgical pro-cesses. Biuletynyn Urzędu Patentowego. 2013;**21**:26. ISSN: 0137-8015

[9] Holtzer M, Dańko R, Dańko J, Kubecki M, Żymankowska-Kumon S, Bobrowski A, Śpiewok W. Ocena szkodliwości materiałów wiążących stosowanych do mas formierskich i rdzeniowych nowej generacji. (Akapit). Kraków: AGH; 2013. ISBN 978-83-63663-69-8

[10] Bobrowski A, Holtzer M, Dańko R, Żymankowska-Kumon S. Analysis of gases emit-ted during a thermal decomposition of the selected phenolic binders. Metallurgia International. 2013;**18**(7):259-261. ISSN 1582-2214

[11] Holtzer M, Kwaśniewska-Królikowska D, Bobrowski A, Dańko R, Grabowska B, Żymankowska-Kumon S, Solarski W. Investigations of a harmful components emission from moulding sands with bentonite and lustrous carbon carriers when in contact with liquid metals. Przegląd Odlewnictwa. 2012;**62**(3-4):124-132

[12] Zych J, Mocek J, Snopkiewicz T. Gas generation properties of materials used in the sand mould technology – Modified research method. Archives of Foundry Engineering. 2014;**14**(3/2014):105-109

[13] Zych J, Mocek J. Kinetics of gas emissions from moulding and core sands, gasification patterns and protective coatings – The new investigation method. Manufacturing Technology. 2017;**17**(1):126-131. ISSN 1213-2489

[14] Zych J, Mocek J. Protective coatings of moulds and cores – Additional sources of gases and potential causes of surface defects. Przegląd Odlewnictwa, Kraków. 2017;**67**(3-4): 90-95. ISSN 0033-2275

[15] Mocek J, Zych J. Kinetics of gas emission from heated moulding sands together with the on-line assessment of H_2 and O_2 fractions - new investigation method. Archives of Foundry Engineering. 2016;**16**(4):79-84. ISSN 1897-3310

[16] Mocek J, Zych J, Krubnik Ł. Casting of Al-Si alloys in the full mould technology process conditions. International Conference on Science and Technology. Akapit – Kraków; 2014. pp. 73-84. ISBN: 978-83-63663-45-2

Chapter 3

Diels-Alder Chemistry to Develop Self-Healing Epoxy Resins and Composites Thereof

Stefania Dello Iacono, Alfonso Martone and
Eugenio Amendola

Additional information is available at the end of the chapter

http://dx.doi.org/10.5772/intechopen.81360

Abstract

Self-healing polymers are a class of smart materials able to recover after sustaining damage. A family of thermosetting epoxy resins, containing Diels-Alder (DA) adducts in the epoxy precursor backbone, has been prepared and characterized. The DA adducts can be reversibly cleaved and reformed under the appropriate thermal conditions, and this feature has been exploited to produce intrinsically self-repairing materials. This chapter focuses on the effects of different structural features, such as average number of cross-linking functionality and molecular mobility of epoxy precursors, on the efficiency of healing process. High cross-linking density and molecular stiffness improve mechanical performances, such as elastic modulus and glass transition temperature, and allow fabrication of self-healing fiber-reinforced composites by conventional manufacturing technologies. Within this chapter, the molecular design, the preparation, and the evaluation of properties of self-healing epoxy and its composites have been discussed.

Keywords: thermo-reversible network, fracture toughness, self-healing, intrinsic healing, sustainable materials

1. Introduction

High-performance polymeric materials used in the automotive, aerospace, and space industries are progressively replacing metals as structural materials. However, their performances, such as mechanical properties, fatigue life, and esthetic features are usually deteriorated by severe in-service loads and/or environmental conditions. Therefore, the integration of recovery damage capability represents a major challenge for the next generation of technopolymers. [1]

Soon after the landmark work by White and Sottos [2] demonstrating the concept of self-healing features in a man-made material, the academic and industrial scientists recognized the ability of this new class of polymers to mitigate the effects of local damages in order to restore mechanical or functional properties and to postpone catastrophic failure of the whole structure. Two primary healing methodologies in polymers focus either on the incorporation of encapsulated healing agents in the polymer bulk or on the introduction of dynamic bonds in the macromolecular backbone. The former mechanism is referred to as extrinsic healing [2–4]: it is autonomic and is limited to a single, or to very few, healing events occurring in the same site. While the latter is referred to as intrinsic healing [5, 6]: it requires an external stimulus for its activation and can be applied multiple times on the same damaged spot. A thermal process is convenient and effective for treatment of polymers with a wide range of sample sizes and treatment durations. As a result, temperature-dependent reversible covalent cross-linking of polymers or block copolymers, such as DA-based materials, represents an effective method for the implementation of intrinsic self-healing into functional materials [5–8].

Self-healing thermosets potentially offer increased safety and durability of artifacts produced thereof, and remendable materials are particularly desirable for severe load-bearing applications in which repair and maintenance are costly and safety is concerned.

In addition, costs for material development and production would greatly benefit from the possibility to combine new self-healing materials with conventional resin and from the compatibility with current processing techniques. Nevertheless, the heterogeneity of people involved led to many strategies for improving the durability of existing materials or the synthesis of brand-new polymeric systems. [1] The layout of an efficient strategy to achieve healing efficiency and balance it with mechanical performances exhibited by this new class of materials is a complex task.

This chapter will describe the results we achieved in developing a self-healing epoxy system based on DA reaction. At first, design parameters of an epoxy adduct able to induce mendability will be discussed. Following, techniques to assess the healing capability of a polymeric material are described. Finally, the effort to fabricate composite structural material made by Diels-Alder (DA) thermoset will be discussed [8–10].

2. Molecular design for self-healing

Intrinsic healing mechanism relies on complex chemistry, and its development affords several combined advantages. Therefore, following recent interest of scientific community on this topic, development of chemical pathways leading to self-healing strategy will be discussed.

Several reversible bonds have been used to achieve self-mending functionality. Hydrogen bonds, Van der Waals forces, and electrostatic interactions in polymeric ionomers [11] are claimed to explain self-repair features in supramolecular structures, while covalent disulfide bridges [12], ester linkages [13], alkoxyamine moieties [14], and Diels-Alder bonds [15, 16] account for damage recovery in cross-linked structures. Among them, Diels-Alder chemistry has been widely adopted because of its simplicity, high efficiency, and repeatability through only the application of heat.

Diels-Alder chemistry was first described by Otto Diels and Kurt Alder in 1928 [17] and is particularly useful in synthetic organic chemistry as a reliable and clean method for introducing a six-membered DA adducts on a wide range of organic substrates which can be end-capped by reactive functional groups, such as epoxies, acrylates, amines, isocyanates, and hydroxyls. The DA reaction is a thermally reversible cycloaddition between a conjugated diene and a dienophile resulting in a cyclohexene derivative. The cross-linked DA adducts can undergo a cleavage reverse reaction at higher temperatures (rDA) [18, 19]. While several diene-dienophile couples are available for DA reaction, to date, the most investigated precursors are furan/maleimide derivatives. The bond energy of the new C-C σ bonds in DA adducts was evaluated to account to 96.2 kJ/mol [20], while other covalent bond energies are 348 kJ/mol for C-C bonding and 293 kJ/mol for C-N [21]. Since covalent bonds are three to four times stronger than C-C σ bonds formed in DA adducts, cracks are more likely to form and propagate between the new formed bonds in DA adducts. Consequently, available diene and dienophile on the freshly generated surface increase the efficiency of the self-healing method, through the DA recombination. Another notable implication, resulting for the diene-dienophile choice, is the temperature where the self-recovery can be achieved. The range for the rDA reaction of furan/maleimide derivatives is approximately at temperatures higher than 115–120°C, while at lower temperatures the DA recombination is favored.

When considering self-healing thermosets with the DA reaction, the local molecular mobility, displayed during mending cycles, is a key parameter affecting the healing efficiency. Fast and efficient damage recovery and DA bond recombination are achieved via a local and temporary increase of mobility occurring at temperatures higher than polymer glass transition temperature but lower than rDA cleavage temperature. In the case of DA adducts based on the furan/maleimide couple, the processing window can be identified between the polymer's T_g and rDA temperature.

Epoxy resins have been widely used due to their excellent heat resistance, outstanding corrosion protection, high electrical resistivity, and superior mechanical properties. However, cracks might occur as a result of thermal stress and mechanical fatigues during processing and service conditions. Many papers already described epoxy resins containing DA adducts in their backbone [22, 23].

Following a related concept, bifunctional epoxy precursors cross-linked with amines [24] will be described in the present work. The choice to locate the reversible DA bond on the epoxy or on the amine moieties results in structurally equivalent networks, as far as the self-healing phenomena are concerned. Nevertheless, the synthetic path to produce epoxy DA adducts is easier with respect to the preparation of DA amines, due to the added complexity of amine protection/deprotection. For the sake of simplicity, tetrafunctional amines were used as cross-linkers in the following, while the epoxy functionality was varied between two and four, to investigate about the effect of different cross-linking density.

Cross-linking density and conformational stiffness of molecular fragments between adjacent cross-links are the main parameters, which affect properties, such as T_g, mechanical stiffness, and overall molecular mobility. However, accurate experimental evaluation of cross-linking density is often difficult, especially for glassy and rigid polymers. Therefore, the use of an easily defined marker as a measure of cross-linking density is strongly envisaged. In this respect, the average functionality of a mixture of monomers can be defined as the average number of functional groups per monomer molecule for all types of monomer molecules. It is defined by

$$f_{avg} = \frac{\sum N_i f_i}{\sum N_i} \tag{1}$$

where N_i is the number of molecules of monomer i with functionality f_i and the summations are over all the monomers present in the system. As a rule of thumb, reactive mixture with f_{avg} strictly equal to 2 can lead to high molecular weight linear polymer and f_{avg} smaller than 2 results in low molar mass oligomers, while f_{avg} greater than 2 produces branched and cross-linked networks. In our specific case, for a system consisting of 2 mol of di-epoxy and 1 mol of tetrafunctional amine, f_{avg} is 2.67; for a system consisting of 1 mol of tetrafunctional epoxy and 1 mol of tetrafunctional amine, f_{avg} is 4.00. All the intermediate values are achievable by proper adjustment of the ratio between bifunctional versus tetrafunctional epoxy, stoichiometrically balanced by the amine.

In the case of conventional cross-linked networks, f_{avg} is invariant with respect to temperature, and the network cannot flow upon heating. If the network precursors include reversible bonds, such as DA, f_{avg} decreases at high temperatures, after triggering the rDA cleavage reaction. Referring to the bi- and tetrafunctional DA epoxy cross-linked with conventional tetrafunctional amine, the f_{avg} drops to 1.60 in the former case, while the value is 2.67 for the latter one. Obviously, a high cross-linking density hinders molecular mobility: a value of f_{avg} higher than 2 prevents the healing phenomenon, if the condition holds true also in the cleavage state at high temperature.

The ability to adjust the molecular mobility during the healing cycle at a desired preset value is an important molecular design tool, which is helpful in addressing specific application requirements. For example, if the thermosetting DA resin is intended for restoration of small impact damages or micro-delaminations, which are often encountered in the case of barely visible impact damages (BVID) of composite materials, a moderate molecular mobility is required. The sample has to retain its geometry and fiber placement, and long-range viscous flows are detrimental. This target can be achieved tailoring the cleaved state f_{avg} so as its value is close to 2. On the contrary, if thermal recycling of thermoset is sought after, a high extent of molecular mobility is required in the cleaved state to allow materials to flow. While in the first case the overall material properties closely resemble the conventional thermosets, in the latter case, a dynamic thermoplastic-like state is achieved during self-healing. The epoxy mixture formulation can be properly adjusted to suit one of the previous conditions to achieve bespoke molecular mobility, measured by f_{avg}. For this purpose, cross-linking density of the cleaved stage can be preset to different levels, corresponding to reduced molecular mobility, by mixing bifunctional Diels-Alder epoxy with conventional epoxy or mixing a bi- and a tetrafunctional Diels-Alder epoxy. The introduction of conventional epoxy brings the added benefits of overall reduced costs because of the use of a cheaper precursor, while the introduction of tetrafunctional Diels-Alder epoxy benefits the healing efficiency increasing the overall concentration of DA functional groups.

Further degree of freedom can be introduced in the molecular design if detailed precursor structure is considered. In fact, according to **Figure 1(a)**, a symmetrical Diels-Alder precursor, bearing a pair of cleavable dienophile groups, can be considered. But the same structural features can be achieved using a smaller molecule, as shown in **Figure 1(b)**. In the former case, the formation of an unbound dienophile allows higher molecular mobility, increasing therefore the healing efficiency.

The coexistence of a stable and a thermo-reversible polymeric network, required for the development of robust self-healing ability [8, 25], is depicted in **Figure 2**. The hybrid polymer architecture is guaranteed by irreversible cross-links of conventional epoxy (green oval) with tetrafunctional amines (green rectangle) and reversible covalent bonds between

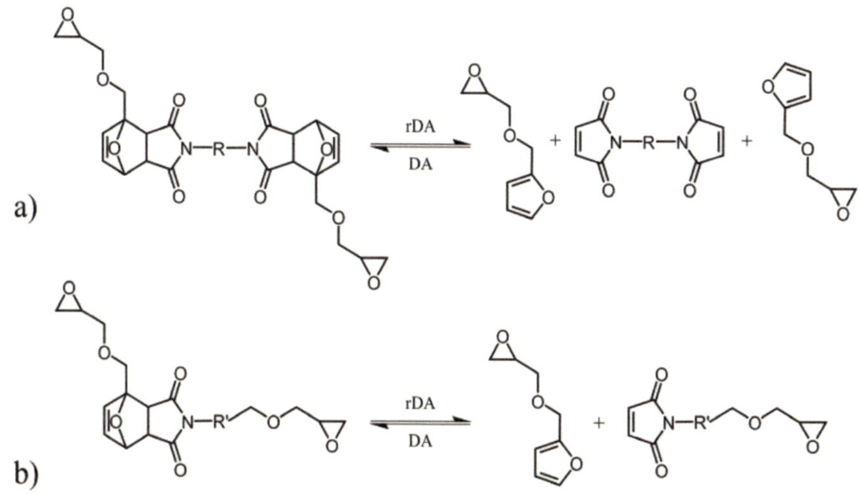

Figure 1. Examples of of Diels-Alder adducts: a) symmetric molecules, b) asymmetric molecules.

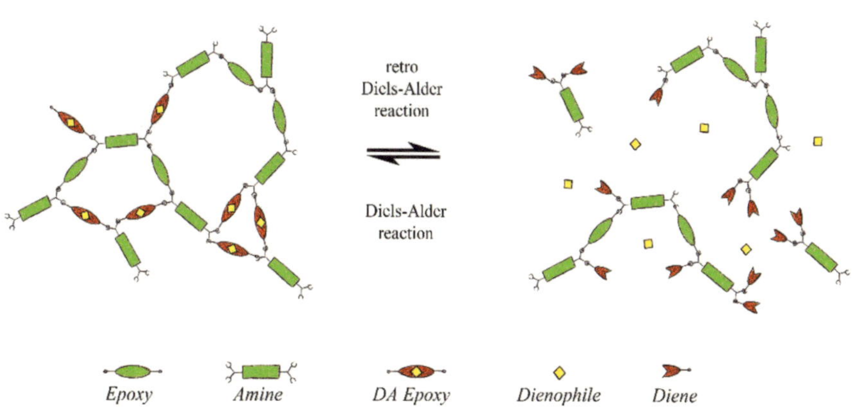

Figure 2. Hybrid network scheme.

furan derivative diene (red pincer) and bismaleimide derivative dienophile (yellow square) of a DA adduct.

The left-hand side of **Figure 2** is representative of the cross-linked structure which prevents viscous flows. At low temperatures, the material behaves like a thermoset. As the temperature is increased above the rDA threshold trigger, the cleavage of epoxy DA activates a higher molecular mobility. Small molecular fragments, depicted in the right-hand side schematic of **Figure 2** as result of retro Diels-Alder reaction, symbolize distinguishing mobile state and viscous flow of thermoplastic material. Therefore, the presence of thermo-reversible chemical bonds switches the material state between thermoset-like behavior at low temperature and thermoplastic-like flow at high temperature.

3. Chemical-physical assessment on self-healing features

The presence of mechanical, thermal, or electrochemical damages results in regions with reduced performances with respect to the surrounding materials and deteriorates the overall response of the component.

The evaluation of self-healing efficiency is a complex task, which is not extensively disciplined yet in test standards or experimental procedures. It can be performed at different dimensional scales, starting from visual inspection or optical microscopy (OM) observation, up to spectroscopic techniques based on molecular interaction, comprising mechanical and dynamical-mechanical characterization [26].

The self-healing mechanism of thermo-reversible epoxy resins relies on the direct and reverse cycloaddition Diels-Alder reaction, which could be monitored by FTIR through the investigation of the spectral band of the C-O-C peak around 1180 cm^{-1} [8, 27]. Similarly, stretching vibrations in Raman spectra of C=C at 1501, 1575, 1585, and 1600 cm^{-1} bands, related to furan/maleimide-based DA adduct, could be used to monitor the progress of the reactions [28]. Due to the system complexity, only the signal at 1501 cm^{-1}, ascribable to C=C stretching vibration of the furan ring [28, 29], is a useful marker.

Since reversible epoxy thermosets exhibit properties of both highly cross-linked epoxy thermosets as well as typical behavior of thermoplastics that soften and flow at elevated temperatures, the study of thermomechanical behavior of self-healing DA system allows the identification of macroscopic behavior related to low-temperature highly cross-linked status or the high-temperature viscous thermoplastic-like condition.

Figure 3 depicts the healing and reshaping cycles of a composite coupon based on self-healing resin. Thermo-reversible bonds affect the molecular mobility [30]; therefore, the investigation of linear viscoelastic behavior is a suitable procedure to assess the self-healing feature of the material.

Rheological tests can be used to discriminate between the solid like behavior and the semi-viscous state of degenerated networks, by measuring storage and loss moduli as a function of temperature [8, 31, 32].

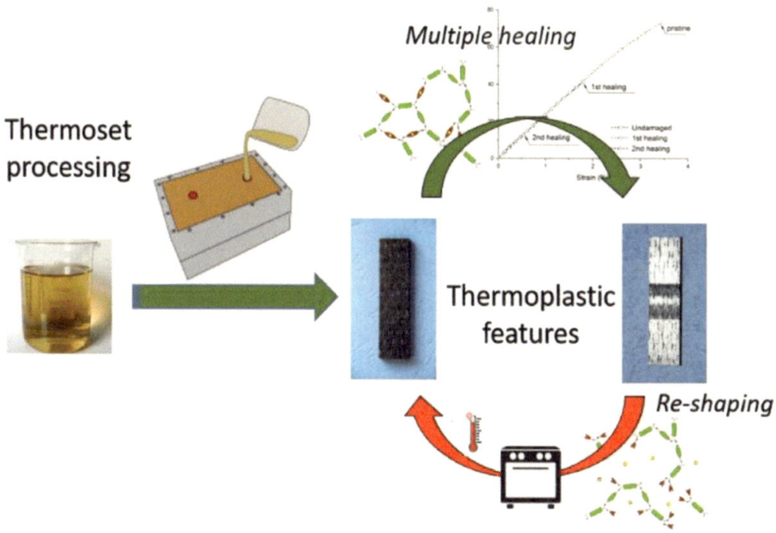

Figure 3. Schematic of the transition between thermoset state and thermoplastic state triggered by temperature.

On the other hand, rheological experiments can effectively complement the mechanical and dynamical-mechanical characterization. Thermo-reversible epoxy resins, similar to conventional ones, exhibit little elongation at break. This characteristic depends on the degree of cross-linking of the polymer chains, and then failure of such systems could be described in terms of linear fracture mechanics [6].

The experimental assessment of the fracture properties relies on the study of short-term monotonical loading.

The failures in polymeric materials are a result of chain scission and structural breakup. Self-healing recovery of damaged polymeric structure prevents or reduces the fracture propagation. When the applied load exceeds the critical fracture stress, the crack grows. If the healing cycles preserve the initial shape and dimensions, the healing efficiency, η, may be calculated as expressed in Eq. (2):

$$\eta = \frac{\sigma_{crit}^{healed}}{\sigma_{crit}^{virgin}} \approx \frac{K_{crit}^{healed}}{K_{crit}^{virgin}} \tag{2}$$

where σ_{crit}^{healed} is the stress required to propagate the crack to a given length in the healed material and where σ_{crit}^{virgin} is the stress required to propagate the crack to approximately the same length in the virgin one. Analogously, K_{crit}^{healed} and K_{crit}^{virgin} are the stress-intensity factors for the healed and the virgin materials, respectively.

However, while the efficiency defined in terms of the applied stress represents an actual measured quantity, the same quantity expressed in terms of the stress-intensity factor represents an estimate using a fracture model. The lack of control inherent in the fracture process precludes a direct comparison.

Generally, the efficiency of the healing process is evaluated by the experimental comparison between the performance of intact and healed material [2]:

$$\eta = \frac{P_{healed}}{P_{initial}} \cdot 100 \tag{3}$$

The subscripts refer to whether the property is measured after healing (*healed)* or before damage occurs (*initial*). In many cases, healing efficiency is defined in terms of the fracture toughness [33] or in terms of material strength [34, 35] or material stiffness [10, 12].

4. Self-healing Diels-Alder epoxy systems

Self-healing feature increases the durability of thermosets by reducing the service costs for high-end applications. Unfortunately, the use of smart polymeric materials raises the issue of a compromise between material performance and integration of self-healing properties. Rigid materials have specific properties, which should remain unchanged by the self-healing chemistry both during use and later to a healing treatment. Such surfaces are usually made of highly cross-linked thermoset polymers; the incorporation of self-healing functionalities into their formulations can be problematic and needs to be investigated.

4.1. Synthesis of epoxy Diels-Alder adducts

Based on the hierarchical criteria developed on Section 2, a family of epoxy precursors was prepared integrating Diels-Alder precursors 2Ph2Epo and 2Ph4Epo cross-linked with tetrafunctional amines (**Table 1**).

The two precursors differ in number of epoxy groups, keeping central furan/maleimide DA adduct fixed. Structural formulae of epoxy precursors are shown in **Table 1**. The 2Ph2Epo is characterized by the presence of two oxirane rings and two Diels-Alder adducts. The introduction of two additional functional groups results in 2Ph4Epo.

Acronym	Molecular structure	Mw (gmol^{-1})
2Ph2Epo	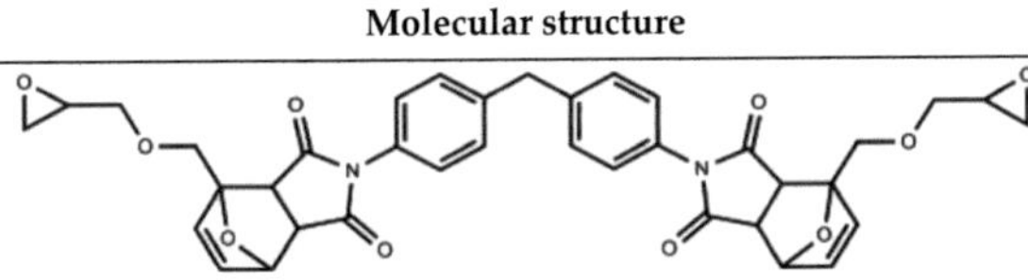	666.68
2Ph4Epo	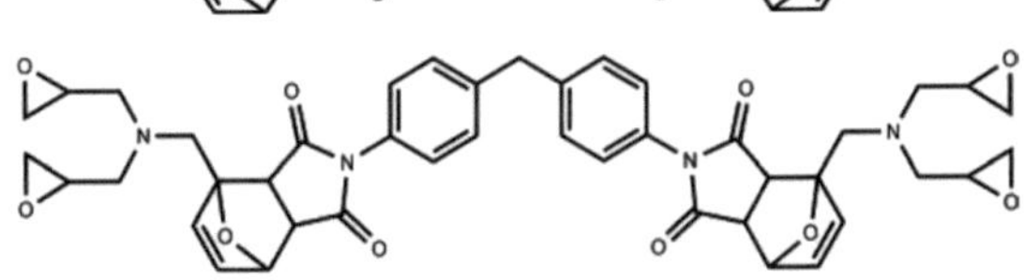	776.84

Table 1. Diels-Alder epoxy precursors.

To induce the self-healing capability, two identical dienes (furfural derivatives) were capped on a bismaleimide dienophiles, resulting in a symmetrical epoxy compound containing two Diels-Alder adducts. The synthesis of 2Ph2Epo has already been described by the authors in [24]. ^{1}H-NMR spectrum, recorded in d_6-DMSO with Bruker Avance 400, points out the appearance of characteristic peak of DA adduct (δ~5.5), because two carbon atoms change from sp^2 to sp^3 hybridization (double peak slightly split for the formation of the two stereoisomers, endo and exo, with the former overwhelming the latter).

Analogously, the use of furan derivative with a pair of oxirane rings results in the preparation of 2Ph4Epo adduct. Also in this case, the ^{1}H-NMR spectrum confirmed the accuracy of structure reported in **Table 1**.

4.2. Cross-linked sample preparation

The Diels-Alder epoxies 2Ph2Epo and 2Ph4Epo and their mixture with DGEBA were cross-linked using stoichiometrically balanced DDM and Jeff500 as curing agents (**Figure 4**) at 90°C for 24 hours.

Suitable samples of self-healing epoxy resin have been developed by application of the criteria shown in Section 2. To vouch for a complete physical-chemical and technological compatibility with conventional epoxy system and production technology used in the field of epoxy resin and composites thereof, a mixture of commercial tetrafunctional amines was used as curing agent. The mixture of two different amines was used to finely tune the glass transition temperature of the self-healing resin in the range of 90°C and achieve full cure at temperatures below rDA reaction without incurring the temporary scission of epoxy precursor. The cross-linking density and f_{avg} during the healing process were controlled by the proper ratio between the functional DA epoxy and a commercial epoxy resin.

Composition of all samples is reported in **Table 2**.

Samples reported in **Table 2** were cured at 90°C for 24 hours. DSC confirmed the complete conversion of cross-linking reaction by the absence of residual reactivity.

4.3. Self-healing properties

As already discussed, the self-healing capability of small fractures and BVID is related to local molecular mobility, temporarily activated by temperature increase. The observation of superficial scratches and their recovery is a generally accepted technique for the assessment of self-healing features. For this task, a controlled mark has been produced by sharp scalpel

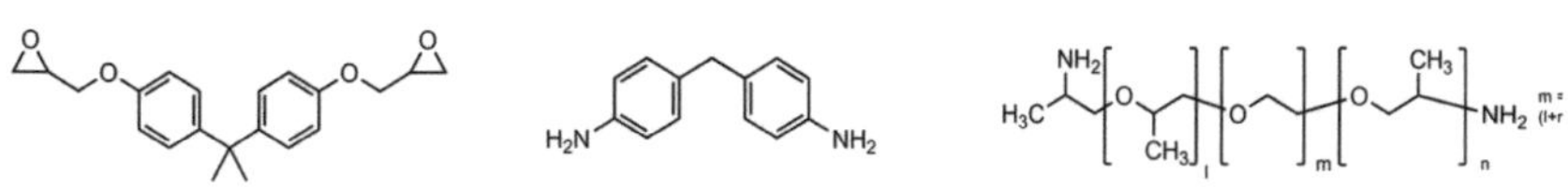

Figure 4. Molecular structures of DGEBA, DDM, and Jeff500.

Acronym	DA epoxy (mol)	DGEBA (mol)	DDM (mol)	Jeff500 (mol)
DGEBA100	–	1.00	0.30	0.20
2Ph2Epo100	1.00	–	0.30	0.20
2Ph2Epo65	0.65	0.35	0.30	0.20
2Ph4Epo100	1.00	–	0.60	0.40
DGEBA100	–	1.00	0.30	0.20

Table 2. Cross-linked sample composition.

and observed by optical microscopy (Olympus BX 51 M), applying suitable thermal stimulus by means of Linkam THM600 hot stage.

As expected, microscopy observation performed on cross-linked DGEBA resin cured with a mixture of 60/40 mol/mol of DDM and Jeff500 amines (DGEBA100, T_g = 90°C) evidenced the lack of scratch recovery even at 140°C (**Figure 5**), well above sample T_g.

Only a minor modification of scratch width can be detected as a result of stress relaxation at temperature higher than resin T_g. Nevertheless, self-healing is hindered by cross-links.

On the other hand, thermal treatment at 120°C for 5 min completely restored the damaged surface of 2Ph2Epo100 sample, as depicted in **Figure 6**. Unfortunately, the high molecular mobility achieved in the activated stage produced sample deformation and viscous flow. In fact, due to the high concentration of reversible bonds, the cross-linking density in the cleaved stage dropped, and materials transformed into viscous thermoplastic.

The occurrence of this phenomenon is not desirable if self-healing materials have to be used for structural application and the molecular mobility has to be reduced, either by introducing a thermally stable epoxy precursor such as DGEBA or by using a tetrafunctional DA precursor.

In the first case, the 2Ph2Epo65 system, containing 65% of DA epoxy and 35% of DGEBA (**Table 2**), was prepared, with T_g of 90°C. Self-healing capability of 2Ph2Epo65 was confirmed even after reduction of thermo-reversible bond concentration. The increased network stability preserved the sample shape and dimension still allowing self-healing phenomenon, as

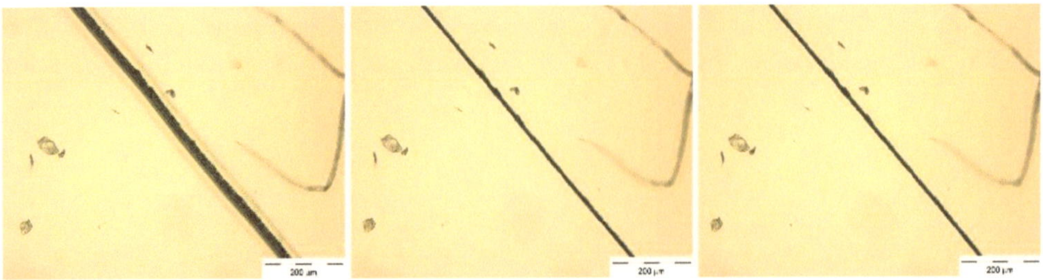

Figure 5. Scratch recovery for DGEBA100, 10 × magnification. From left to right: at room temperature, at 140°C, after additional annealing for 20 min at 140°C.

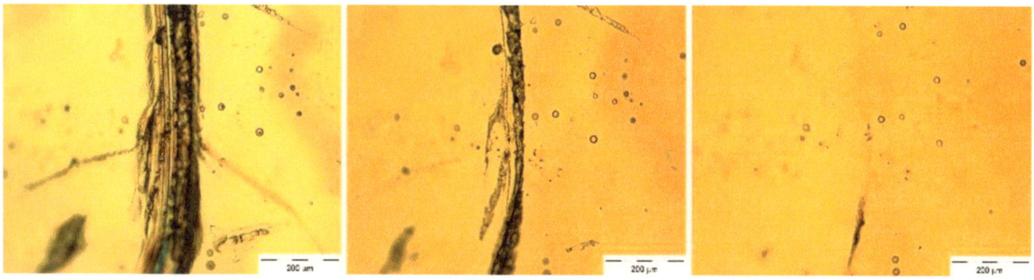

Figure 6. Scratch recovery for 2Ph2Epo100, 10 × magnification. From left to right: at room temperature, at 120°C, after additional annealing for 5 min at 120°C.

reported in **Figure 7**. The scratch completely disappeared after 30 min at 120°C, restoring the pristine surface.

The second approach to prevent viscous flow of materials during the high-temperature stage is to use Diels-Alder epoxy adduct with functionality higher than 2. The presence of four reacting epoxy groups for each precursor molecule (2Ph4Epo) increases the cross-linking density. 2Ph4Epo100 was prepared according to **Table 2** and fully cured at 90°C for 24 hours, reaching a T_g of 95°C. The occurrence of rDA reaction reduces the cross-linking density at high temperature. But the f_{avg} calculated in the cleaved state for tetrafunctional DA epoxy is 2.67 and suggests hindered molecular mobility, as already discussed in Paragraph 2. As a consequence, morphological damages were not recovered for 2Ph4Epo100 (**Figure 8**).

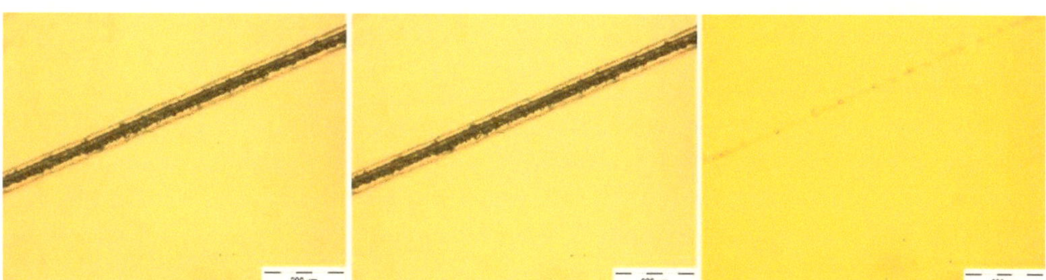

Figure 7. Scratch recovery for 2Ph2Epo65, 10 × magnification. From left to right: at room temperature, at 120°C, after additional annealing for 30 min at 120°C.

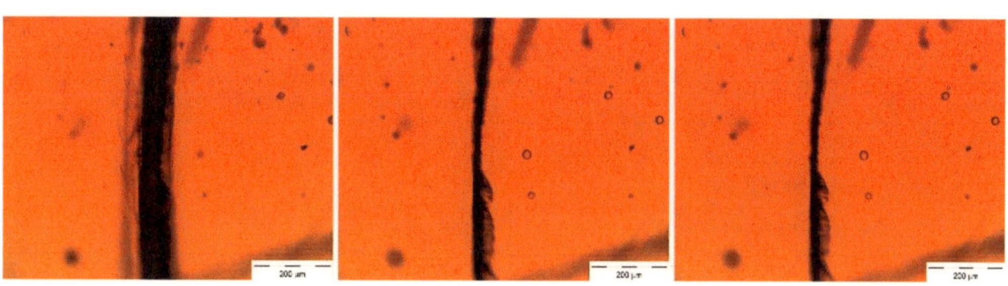

Figure 8. Scratch recovery for DGEBA2Ph4Epo100, 10 × magnification. From left to right: at room temperature, at 120°C, after additional annealing for 20 min at 120°C.

Upon sample breakage, Diels-Alder bonds are preferentially cleaved because they are weaker than other covalent bonds building up the cross-linked network. Therefore, the occurrence of mechanical damage makes diene and dienophile groups available for self-healing on the fracture surface. But molecular backbone in the close proximity of damage remains unaffected, and the overall material stiffness prevents an efficient fracture healing. For this reason, further thermal treatment at 120°C is applied to complete cleavage of DA bonds and to maximize molecular mobility. Physical healing, due to diffusion of molecular fragments in the activated stage, allows the fracture edges recombination. After the first step, a further annealing at 90°C is required to restore the pristine cross-linking density and mechanical properties by direct Diels-Alder reaction.

The proposed healing mechanism is validated by analysis of micro-mechanical tests, performed by Micro Materials NanoTest™ Platform. 2Ph2Epo65 properties have been evaluated and compared between the pristine as prepared and after incremental treatments, including morphological healing and structural annealing.

The reduced elastic modulus, *Er*, was calculated based on Eq. (4), taking into account the effect of nonrigid indenter column:

$$E_r = \frac{\sqrt{\pi}\,S}{2\beta\sqrt{A}} = \left[\frac{(1-\upsilon_i^2)}{E_i} + \frac{(1-\upsilon_S^2)}{E_S}\right]^{-1} \tag{4}$$

where A is the contact area, β the geometric constant (1.034 for a Berkovich indenter), and S the unloading stiffness at maximum load. E and ν are the elastic modulus and the Poisson ratio; and the subscripts "i" and "s" refer to the diamond indenter and the specimen, respectively. The E_i is 1140 GPa, the ν_i is 0.07, and the ν_s is 0.35.

All data were corrected for thermal drift and instrument compliance and subsequently analyzed with the Oliver and Pharr method [36]. According to Zheng [37], elastic modulus measured by depth indentation technique overrates the elastic modulus by a factor of 5–20%. Reduced modulus is reported in **Table 3**. After heating at 120°C, required to promote the morphological recombination of scratch edges, the modulus drops down by a factor of 2. In fact, the occurrence of rDA reaction induces the cleavage of specific covalent bonds and reduces the cross-linking density. However, the effect is not permanent. The pristine properties can be recovered by prolonged annealing at 90°C, when DA reaction can lead to network restoration.

Specimen type	Reduced modulus, E_r(GPa)
As prepared	4.80 ± 0.03
Morphological healing: 20′ @ 120°C	2.54 ± 0.05
Structural healing: 20′ @ 120°C + 12h @ 90°C	4.56 ± 0.05

Table 3. Reduced modulus (GPa) of self-healing 2Ph2Epo65 epoxy resin.

5. Crack healing in long fiber composites

The self-healing epoxy 2Ph2Epo65 has been considered for manufacturing a composite plate with the aim to investigate the fracture behavior and to assess the healing efficiency of the system. Interlaminar behavior of the CFRP (composite fiber-reinforced plastic) has been studied by means of shear strength of the laminate.

Shear tests can be conducted on composite laminates following different experimental approaches, depending on the mode of fracture that needs to be assessed. The ability to recover damages after a cohesive failure has been investigated by interlaminar shear strength (ILSS), and mode II fracture loading has been studied by performing the End Notch Failure (ENF).

A composite plate has been manufactured by liquid molding process under vacuum bag; 12 unidirectional layers were laminated to reach a nominal thickness of 3 mm. The presence of a Kapton layer in a bending test (in the case of ENF tests) leads to the mutual sliding of separated parts promoting a mode II failure (shear mode) [38], according to ASTM D7905.

Load versus displacement curves during ENF tests is reported in **Figure 9** and show an initial linear behaviour up to the critical load (nonlinearity load, NL). Above this point, delaminations start and steadily propagate until the maximum load is achieved. Afterwards, unstable delamination growth leads to load decrease.

The first healing treatment allowed to recover the pristine stiffness, with a sample strength decrease. However, during the third load cycle (i.e., after the second healing), a significant stiffness loss is experienced. Different behaviors between neat polymer and laminate should

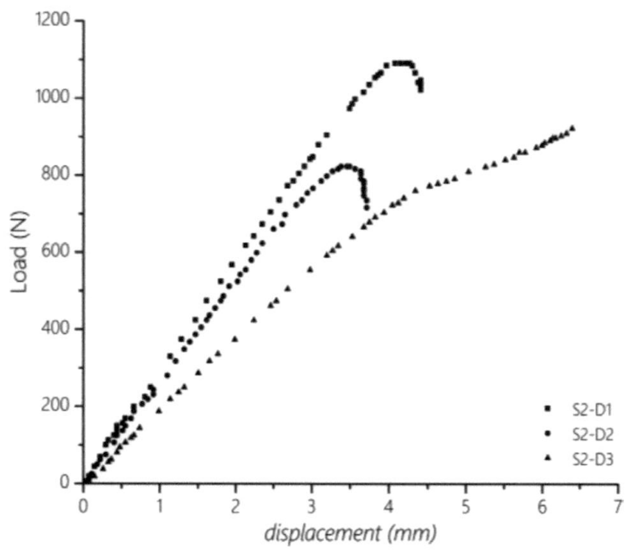

Figure 9. End Notched Failure test.

Cycle	ILSS (MPa)	η (%)	G_{IIc} (J/m²)	η (%)
0	54.1 ± 1.1	100	650 ± 20	100
1	51.4 ± 5.1	95.1	583 ± 13	89.7
2	44.2 ± 4.9	81.7	341 ± 32	52.4

Table 4. Strength recovery after failures.

be related to specific composite features. Delaminations could occur both as effect of a matrix failure and as interface debonding. Moreover, any damage incurring to the reinforcing fibers would reduce material stiffness without chance of recovery.

The critical strain energy release rate should be evaluated as function of the NL load:

$$G_{II_c} = \frac{3\,m\,P^2\,a^2}{2\,B} \tag{5}$$

Fracture toughness was determined using Eq. (5), where *m* is the calibration compliance, *P* is the critical load, *a* is the critical length, and *B* is the specimen width.

Table 4 reports the recovery efficiencies measured as interlaminar critical strength and as critical energy for mode II delaminations. Fracture toughness showed a less effective recovery ability of 52.4% compared to static interlaminar strength recovery of 81.7%.

6. Conclusions

The development of self-healing materials is a very attractive approach to provide long-lasting and efficient protection against micro damages. Great attention from numerous research groups has been paid to polymers, composites, and coatings, which exhibit self-healing behavior at different dimensional scales. Through the chapter, the overall design flow for achieving hybrid epoxy systems containing covalent thermo-reversible bonds and the preparation and evaluation of selected examples were detailed. This paper underlines that the concurring presence of thermo-reversible covalent bonds and high molecular mobility are essential requirements to develop self-healing systems. The most effective structural modification has been pursued by adjustment of the several features: average functionality of reacting precursor mixture and cross-linking density and thermosetting network and concentration of self-healing reversible bonds. Also, the requirements of easy and efficient self-healing were compromised with development of material properties compliant with structural and semi-structural applications.

In particular, possibility to tailor the properties of "dynamic" epoxy resins containing Diels-Alder bonds would allow the development of novel materials, such as reengineered FRP combining the ease of processability typical of thermosets with reworking/recycling capability at the end of life typical of thermoplastics, as an effort to improve environmental sustainability of advanced materials.

Acknowledgements

The authors thank Mrs. Maria Rosaria Marcedula, Mr. Fabio Docimo, and Mr. Mario De Angioletti for their contribution to the experimental setup and testing.

Conflict of interest

The authors declare no conflict.

Author details

Stefania Dello Iacono, Alfonso Martone* and Eugenio Amendola

*Address all correspondence to: alfonso.martone@cnr.it

CNR-IPCB, Institute for Polymers, Composites and Biomaterials, National Research Council of Italy, Naples, Italy

References

[1] Hillewaere XKD, Du Prez FE. Fifteen chemistries for autonomous external self-healing polymers and composites. Progress in Polymer Science. 2015;**49-50**:121-153. DOI: 10.1016/J.PROGPOLYMSCI.2015.04.004

[2] White SR, Sottos NR, Geubelle PH, Moore JS, Kessler MR, Sriram SR, et al. Autonomic healing of polymer composites. Nature. 2001;**409**:794-797. DOI: 10.1038/35057232

[3] Hansen CJ, Wu W, Toohey KS, Sottos NR, White SR, Lewis JA. Self-healing materials with Interpenetrating microvascular networks. Advanced Materials. 2009;**21**:4143-4147. DOI: 10.1002/adma.200900588

[4] Bergman SD, Wudl F. Mendable polymers. Journal of Materials Chemistry. 2008;**18**:41-62. DOI: 10.1039/B713953P

[5] Tian Q, Yuan YC, Rong MZ, Zhang MQ. A thermally remendable epoxy resin. Journal of Materials Chemistry. 2009;**19**:1289. DOI: 10.1039/b811938d

[6] Liu Y-L, Chuo T-W. Self-healing polymers based on thermally reversible Diels–Alder chemistry. Polymer Chemistry. 2013;**4**:2194. DOI: 10.1039/c2py20957h

[7] Toncelli C, De Reus DC, Picchioni F, Broekhuis AA. Properties of reversible Diels-Alder furan/maleimide polymer networks as function of crosslink density. Macromolecular Chemistry and Physics. 2012;**213**:157-165. DOI: 10.1002/macp.201100405

[8] Dello Iacono S, Martone A, Pastore A, Filippone G, Acierno D, Zarrelli M, et al. Thermally activated multiple self-healing diels-alder epoxy system. Polymer Engineering and Science. 2017;**57**:674-679. DOI: 10.1002/pen.24570

[9] Chen X, Wudl F, Mal AK, Shen H, Nutt SR. New thermally remendable highly cross-linked polymeric materials. Macromolecules. 2003;**36**:1802-1807. DOI: 10.1021/ma0210675

[10] Dello Iacono S, Martone A, Filippone G, Acierno D, Zarrelli M, Giordano M, et al. Insight on mendable resin made by combining Diels-Alder epoxy adducts with DGEBA. AIP Conference Proceedings. 2016;**1736**:1. DOI: 10.1063/1.4949650

[11] Varley RJ, van der Zwaag S. Autonomous damage initiated healing in a thermo-responsive ionomer. Polymer International. 2010;**59**:n/a-n/a. DOI: 10.1002/pi.2841

[12] Amamoto Y, Otsuka H, Takahara A, Matyjaszewski K. Self-healing of covalently cross-linked polymers by reshuffling thiuram disulfide moieties in air under visible light. Advanced Materials. 2012;**24**:3975-3980. DOI: 10.1002/adma.201201928

[13] Montarnal D, Capelot M, Tournilhac F, Leibler L. Silica-like malleable materials from permanent organic networks. Science. 2011;**334**:965-968. DOI: 10.1126/science.1212648

[14] Yoshifumi A, Yuji H, Yasuhiro M, Hideyuki O, Atsushi T. Programmed thermodynamic formation and structure analysis of star-like nanogels with core cross-linked by thermally exchangeable dynamic covalent bonds. Journal of the American Chemical Society 2007;**129**(43):13298-13304. DOI: 10.1021/JA075447N

[15] Rickborn B. The Retro-Diels-Alder reaction part II. Dienophiles with one or more heteroatom. In: Organic Reactions. Hoboken, NJ, USA: John Wiley & Sons, Inc.; 1998. pp. 223-629. DOI: 10.1002/0471264180.or053.02

[16] Rickborn B. The Retro-Diels-Alder reaction part I. C-C dienophiles. In: Organic Reactions. Hoboken, NJ, USA: John Wiley & Sons, Inc.; 1998. pp. 1-393. DOI: 10.1002/0471264180.or052.01

[17] Kloetzel MC. The Diels-Alder reaction with maleic anhydride. In: Organic Reactions. Hoboken, NJ, USA: John Wiley & Sons, Inc.; 2011. pp. 1-59. DOI: 10.1002/0471264180.or004.01

[18] Tasdelen MA. Diels–Alder "click" reactions: Recent applications in polymer and material science. Polymer Chemistry. 2011;**2**:2133. DOI: 10.1039/c1py00041a

[19] Sanyal A. Diels-Alder cycloaddition-cycloreversion: A powerful combo in materials design. Macromolecular Chemistry and Physics. 2010;**211**:1417-1425. DOI: 10.1002/macp.201000108

[20] Chen X, a Dam M, Ono K, Mal A, Shen H, Nutt SR, et al. A thermally re-mendable cross-linked polymeric material. Science. 2002;**295**:1698-1702. DOI: 10.1126/science.1065879

[21] Reger DL, Goode SR, Ball DW. Chemistry: Principles and Practice. CA, USA: Brooks/Cole Pub Co; 2009

[22] Peterson AM, Jensen RE, Palmese GR. Thermoreversible and remendable glass—polymer interface for fiber-reinforced composites. Composites Science and Technology. 2011;**71**: 586-592. DOI: 10.1016/j.compscitech.2010.11.022

[23] Liu Y-L, Hsieh C-Y. Crosslinked epoxy materials exhibiting thermal remendablility and removability from multifunctional maleimide and furan compounds. Journal of Polymer Science, Part A: Polymer Chemistry. 2006;**44**:905-913. DOI: 10.1002/pola.21184

[24] Amendola E, Dello Iacono S, Pastore A, Curcio M, Iadonisi A. Epoxy thermosets with self-healing ability. Journal of Materials Science and Chemical Engineering. 2015;**3**: 162-167. DOI: 10.4236/msce.2015.37022

[25] Garcia SJ. Effect of polymer architecture on the intrinsic self-healing character of polymers. European Polymer Journal. 2014;**53**:118-125. DOI: 10.1016/J.EURPOLYMJ.2014.01.026

[26] Bekas DG, Tsirka K, Baltzis D, Paipetis AS. Self-healing materials: A review of advances in materials, evaluation, characterization and monitoring techniques. Composites. Part B, Engineering. 2016;**87**:92-119. DOI: 10.1016/J.COMPOSITESB.2015.09.057

[27] Araya-Hermosilla R, Broekhuis AA, Picchioni F. Reversible polymer networks containing covalent and hydrogen bonding interactions. European Polymer Journal. 2014;**50**: 127-134. DOI: 10.1016/J.EURPOLYMJ.2013.10.014

[28] Geitner R, Kötteritzsch J, Siegmann M, Bocklitz TW, Hager MD, Schubert US, et al. Two-dimensional Raman correlation spectroscopy reveals molecular structural changes during temperature-induced self-healing in polymers based on the Diels–Alder reaction. Physical Chemistry Chemical Physics. 2015;**17**:22587-22595. DOI: 10.1039/C5CP02151K

[29] Strandman-long L, Murto J. Furfuryl alcohol-III. Infrared, matrix infrared and Raman spectra and ab initio and normal coordinate calculations. Spectrochimica Acta Part A: Molecular Spectroscopy. 1981;**37**(8):643-653. ISSN: 0584-8539. https://doi.org/10.1016/0584-8539(81)80063-3

[30] Kavitha AA, Singha NK. "Click chemistry" in tailor-made polymethacrylates bearing reactive furfuryl functionality: A new class of self-healing polymeric material. ACS Applied Materials & Interfaces. 2009;**1**:1427-1436. DOI: 10.1021/am900124c

[31] Scheltjens G, Diaz MM, Brancart J, Van Assche G, Van Mele B. A self-healing polymer network based on reversible covalent bonding. Reactive and Functional Polymers. 2013;**73**:413-420. DOI: 10.1016/J.REACTFUNCTPOLYM.2012.06.017

[32] Turkenburg DH, Fischer HR. Diels-Alder based, thermo-reversible cross-linked epoxies for use in self-healing composites. Polymer (Guildford). 2015;**79**:187-194. DOI: 10.1016/J.POLYMER.2015.10.031

[33] Brown EN, Sottos NR, White SR. Fracture testing of a self-healing polymer composite. Experimental Mechanics. 2002;**42**:372-379. DOI: 10.1007/BF02412141

[34] Canadell J, Goossens H, Klumperman B. Self-healing materials based on disulfide links. Macromolecules. 2011;**44**:2536-2541. DOI: 10.1021/ma2001492

[35] Zhang H, Xia H, Zhao Y. Poly(vinyl alcohol) hydrogel can autonomously self-heal. ACS Macro Letters. 2012;**1**:1233-1236. DOI: 10.1021/mz300451r

[36] Oliver WC, Pharr GM. An improved technique for determining hardness and elastic modulus using load and displacement sensing indentation experiments. Journal of Materials Research. 2011;**7**:1564-1583. DOI: 10.1557/JMR.1992.1564

[37] Zheng S, Ashcroft IA. A depth sensing indentation study of the hardness and modulus of adhesives. International Journal of Adhesion and Adhesives. 2005;**25**:67-76. DOI: 10.1016/j.ijadhadh.2004.02.004

[38] Fierro G-PM, Pinto F, Iacono SD, Martone A, Amendola E, Meo M. Monitoring of self-healing composites: A nonlinear ultrasound approach. Smart Materials and Structures. 2017;**26**:115015 (17pp). DOI: 10.1088/1361-665X/aa89a8

Chapter 4

Design and Synthesis of Bio-Based Benzoxazines

Acerina Trejo-Machin, Laura Puchot and
Pierre Verge

Additional information is available at the end of the chapter

http://dx.doi.org/10.5772/intechopen.76104

Abstract

Polybenzoxazine (PBz) resins are a new type of high-performance synthetic resins that are attractive alternatives to traditional resins. Their properties include near-zero shrinkage upon polymerization, fast evolution of mechanical properties with the conversion, glass transition temperatures much higher than curing temperatures, low water absorption, and excellent dielectrical and mechanical properties. The development of polybenzoxazines has always been linked to petro-based feedstocks, but for the last 5 years, the number of studies related to bio-based benzoxazines is exploding as a consequence of the versatility of the design of the chemical structure of their monomers. Benzoxazine (Bz) monomers are subjected to cationic ring-opening polymerization (ROP), activated by a thermal treatment in the range of 160–250°C. In addition, Bz synthesis promotes the use of naturally occurring phenolic compounds instead of petroleum-based ones to develop high-performance materials from renewable resources and to fit to REACH restrictions. For this purpose, vanillin, eugenol, and cardanol are examples of bio-phenols bridged with several kinds of aromatic and aliphatic diamines. In this chapter, the synthesis and the properties of di-functional benzoxazine monomers prepared from naturally occurring phenolic compounds are reviewed. Symmetric and asymmetric monomers will be detailed. The last part of the chapter is dedicated to the use of bio-phenols to functionalize polymers and to provide benzoxazine functional groups.

Keywords: benzoxazine, bio-based, cardanol, vanillin, phloretic acid

1. Introduction

Traditional thermosets resins are well known for their high-performance properties [1]. However, these multicomponent thermosets can exhibit shortcomings such as high brittleness, as well as the formation of by-products during curing. They are bi-component materials and their short shelf life prevents any long-time storage [2, 3]. Besides, the raw materials used for the production of the resins are today largely petroleum derived and are increasingly subject to European regulation on registration, evaluation, authorisation and restriction of chemicals (REACH) restrictions. The current concern about the use of petroleum-based materials along with new European restrictions has encouraged researchers to develop sustainable alternatives. Among them, benzoxazine are mono-component resins that have recently emerged as a new class of thermosets and are under continuous development [4].

Polybenzoxazines (PBz) have a wide range of properties to be tailored to a wide range of applications [5]. It makes them attractive alternatives to current applications. In addition to properties similar to phenolic or epoxy resins, PBz have unique features compared to other commodity polymers. They include near-zero shrinkage upon polymerization [6], fast evolution of mechanical properties with the conversion [7], one of the highest char yield among processable resins [8], glass transition temperatures much higher than curing temperatures [8], low water absorption [9], and excellent dielectrical and mechanical properties. Strong acid catalysts are not required for their curing [10]. PBz have similar and even lower costs than epoxy and bismaleimides resins [4]. Due to these outstanding properties and their intrinsically rich molecular design flexibility allowing good compatibility with other active structures, PBz have been promoted, over epoxy and traditional phenolic resins, as better candidates for the application of corrosion protection [11–15]. Indeed, their chemical structure can be easily tailored to meet the requirements of various conditions. Nevertheless, the development of polybenzoxazines has always been linked to petro-based feedstocks while the materials synthesized from petroleum derivatives are becoming more and more a concern for the environment and their use in the industry is thus strongly impeded. In consequence, for the last 5 years, the number of studies related to bio-based benzoxazines is exploding as a consequence of the versatility of the design of the chemical structure of their monomers [16]. Some applications of bio-based polybenzoxazine anticorrosive coatings have already been reported, confirming the great interest in the development of new bio-based PBz with promising physical and mechanical properties for industrial applications [17].

In the following paragraphs, we have chosen to only discuss difunctional benzoxazine monomers as they are more suitable for the elaboration of high-performance materials than monofunctional ones. However, more details can be obtained from the reading of the two following books dedicated to benzoxazine science: Handbook of Benzoxazine Resins [18] and Advanced and Emerging Polybenzoxazine Science and Technology [19].

2. Synthesis and polymerization of benzoxazine

2.1. Synthesis of benzoxazine monomers

Benzoxazine (Bz) resins, typically 1,3-benzoxazines, can be synthesized through a Mannich-like condensation of phenolic derivatives, formaldehyde, and primary amines, followed by a ring-closure process on the phenolic derivative (**Figure 1**) [4].

The general chemical structure of the different kind of benzoxazine monomers is depicted on **Figure 2**. This scheme highlights the great versatility in the design of benzoxazine monomers, gathered in different classes depending on their functionality.

First, there are mono-benzoxazine monomers (m-Bz) which are monomers composed of a single benzoxazine group (**Figure 2a**), in opposition to di-benzoxazine monomers (di-Bz), composed of at least two benzoxazine groups. Symmetric di-Bz monomers can be prepared either from a diphenol (class A, **Figure 2b**) or a diamine (class B, **Figure 2b**). Asymmetric di-Bz can also be obtained (**Figure 2d**) [10]. Finally, multifunctional amines or phenolic derivatives can be used to synthesize polymers bearing multiple benzoxazine groups. Di-phenolic compounds and diamines can also be combined to yield a linear polymer having benzoxazine rings in the main chain. These two last categories are gathered in class C (**Figure 2e**) [4].

Depending on the reagents, temperature, time, solvent, and synthetic procedure, the synthesis of the Bz monomer yields from 70 to 90%. The synthesis can be carried out with or without solvent [20]. This approach is feasible when the mixture of the reactants is liquid or in the molten state at working temperature [21].

2.2. Benzoxazine polymerization

Benzoxazine monomers are subjected to cationic ring-opening polymerization (ROP), activated by a thermal treatment in the range of 160–250°C (**Figure 3**) [20]. The ROP takes place upon heating due to the small amount of impurities generally found in the monomers, such as phenolic raw materials or benzoxazine oligomers. The polymerization is then auto-catalyzed by the formation of phenolic compounds [20].

Mono-functional Bz commonly leads upon polymerization to a linear and low-molecular-weight polymer from few hundreds to some thousands g/mol [22]. Nonetheless, if the aromatic ring is reactive enough, *that is* with available *para* positions for instance, a mono-functional Bz can lead to a cross-linked polymer.

Systematically difunctional or polyfunctional benzoxazine monomers lead to cross-linked structures due to their higher functionality [20]. Thus, difunctional Bz monomers are preferred for the elaboration of high-performance PBz materials.

Although the polymerization of benzoxazine monomers is thermally activated and auto-catalyzed, it requires high temperatures and relatively long reaction time for the complete polymerization. As a result, the use of some initiators or catalysts to accelerate the polymerization

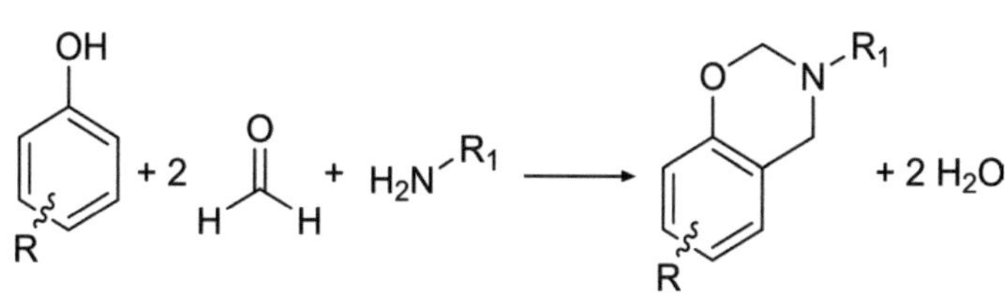

Figure 1. Synthesis of benzoxazine monomer.

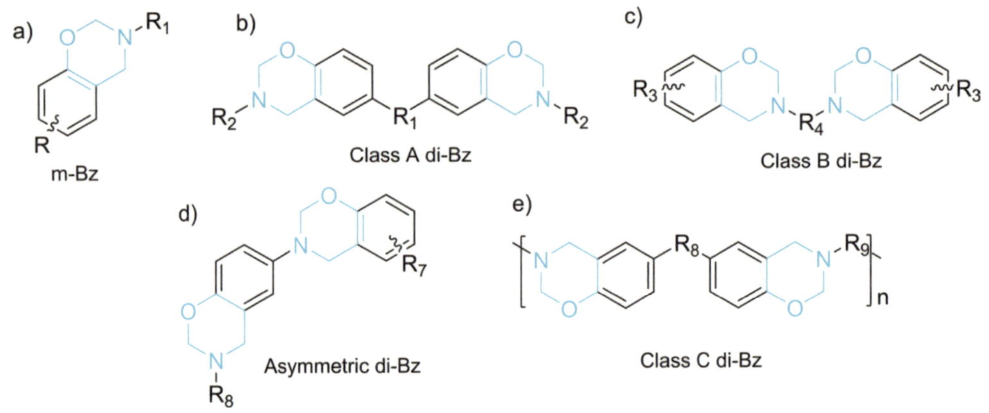

Figure 2. Schematic representation of benzoxazine monomers. Blue labelled rings in the structures correspond to the benzoxazine ring.

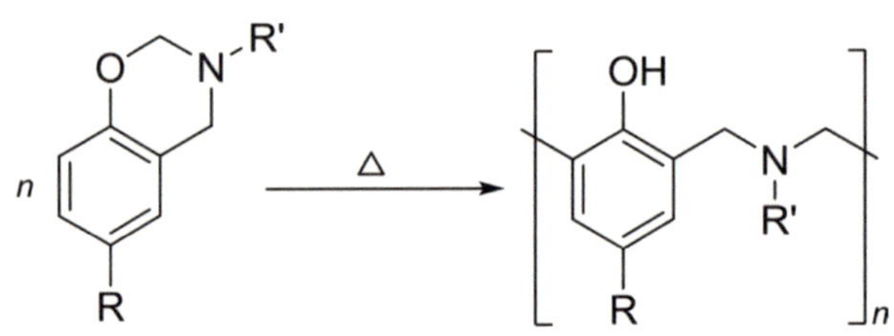

Figure 3. Ring-opening polymerization of benzoxazines.

or to trigger it to lower temperatures is investigated [23]. Most of effective catalysts reported are acidic catalysts, as ROP is a cationic process. Wang et al. [24] studied the use of acidic catalyst for reducing the curing temperature of bisphenol A-aniline–based Bz monomers (BA-a). The polymerization of (BA-a) monomers with 5 mol % of several catalysts was carried out at room temperature during 20 h. The thermal properties of the Bz monomers cured in the presence of Lewis acids such as PCl_5, PCl_3, $POCl_3$, and $TiCl_4$ showed glass transition temperature (Tg) higher than 200°C.

Various phenolic compounds derived from the opening of the oxazine ring or benzoxazine oligomers catalyze benzoxazine ring-opening polymerization reactions, resulting in the autocatalyzed nature of benzoxazine polymerization. In addition, the presence of phenolic structures with free *ortho* positions has a catalytic effect on the curing reaction. Reaction induction time is then decreased, while the reaction rate is enhanced [25]. The benzoxazine polymerization can also be catalyzed by the use of strong acids and/or carboxylic acids. In summary, a broad variety of catalysts may be used for the ring-opening polymerization of Bz monomers at moderate temperature.

Small amount of monocarboxylic monomer induced a reduction of about 20–30°C of the polymerization temperature. When comparing the Tg of the PBz materials, it could be observed that materials obtained from carboxylic-containing Bz monomers tend to have higher Tgs. These results were most probably due to higher cross-linking due to hydrogen bonding with increasing content of carboxylic acid functions.

Specific functional groups have a strong impact on the thermal ROP activation as well as on the mechanical properties of the resultant PBz. In addition, class B Bz monomers promote the use of naturally occurring phenolic compounds instead of petroleum-based ones to develop high-performance materials from renewable resources and to fit to REACH restrictions. For this purpose, vanillin [26], eugenol [27, 28], and cardanol [29–31] have been bridged with several kinds of aromatic and aliphatic diamines. This is discussed in the following part.

3. Bio-based symmetric benzoxazines

3.1. From cardanol

Growing interest has arisen toward the synthesis of Bz monomers stemming from cardanol. Indeed, the particular chemical structure of cardanol appears as a clear asset for the synthesis of PBz materials. Cardanol displays a similar reactivity as phenol through the presence of the hydroxyl group. Furthermore, the long alkyl chain in *meta* position imparts hydrophobicity and flexibility, through internal plasticization of the alkyl chain, to the usually brittle PBz materials. Moreover, the *ortho* and *para* positions of cardanol are available for the synthesis of Bz monomers and their polymerization, respectively.

The chemical structures as well as the thermal properties of cardanol-based di-Bz monomers and their corresponding materials are resumed in **Table 1**.

	Bz monomer	T_m (°C)	T_p (°C)	$T_{5\%}$ (°C)	Ref.
(1)	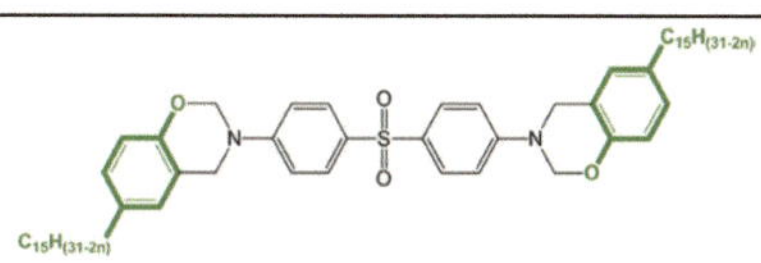 Cardanol/4,4′-diamino diphenyl sulfone (dds)	–	251	299	[32] [33]
(2)	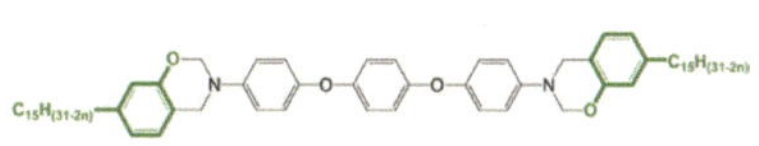 Cardanol/ (4-(4-aminophenoxy)phenyl)ether (44 appe)	114	233	348	[32] [33]
(3)	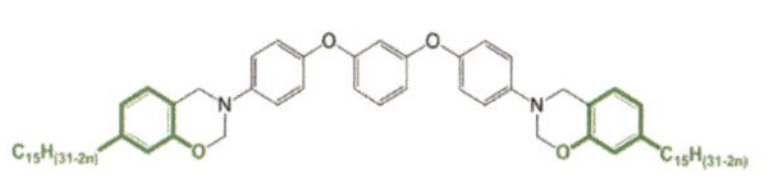 Cardanol/bis-(3-(4-aminophenoxy)phenyl)ether (34 appe)	70	267	354	[32] [33]
(4)	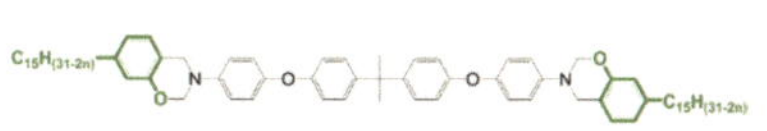 Cardanol/2,2-bis(4-(4-aminophenoxy)phenyl)propane (appp)	–	263	323	[32] [33]

Table 1. Thermal properties of cardanol-based Bz monomers, synthetized from diamines, and $T_{5\%}$ of their corresponding PBz materials.

Almost all the reported cardanol-based Bz monomers synthesized with short amines did not display a melting endotherm. This phenomenon could be attributed to the steric hindrance generated by the C15 alkyl side chain of cardanol that prevents crystallization. However, it can be noticed that high melting temperatures were observed for cardanol-based di-Bz monomers with rigid aromatic diamine as bridging groups, such as bis-(3-(4-aminophenoxy)phenyl)ether or (4-(4-aminophenoxy)phenyl)ether (Tm = 70 and 114°C, respectively) (**Table 1**, lines 2 and 3).

It is noteworthy the thermo-mechanical properties of these PBz materials were almost never characterized by DMA analysis, as it is very difficult to obtain self-supported materials due to the steric hindrance of the alkyl side chain, yielding low-molecular-weight polymers. Consequently, cardanol-based Bz monomers displayed wide processing windows owing to their low melting temperatures due to the steric hindrance brought by the C15 alkyl chain. However, the low cross-linking density and Tg of the corresponding PBz materials, as well as the high polymerization temperatures, are drawbacks for the elaboration of high-performance thermosets and emphasize the use of cardanol-based di-Bz as processing aid (*i.e.*, reactive diluent) for other benzoxazines [34].

3.2. From vanillin

Vanillin is a phenolic compound with a *para* formyl group and *ortho* methoxy group, issued from vanilla seedpod. Vanillin can also be obtained industrially from the processing of lignin [35, 36]. Its use as a precursor for the synthesis of bio-based Bz monomers was recently studied [37].

The formyl group contained in the vanillin compound is of great interest for the synthesis of bio-based Bz monomers. Indeed, with the appropriate stoichiometric amount of reagents and through a convenient order of their introduction and reaction, the aldehyde function is not consumed during the synthesis. This function is thus considered as an additional reactive group on the Bz monomer, able to further react with other chemical compounds or to increase the material cross-linking density.

The chemical structures as well as the thermal properties of vanillin-based di-Bz monomers and their corresponding materials, reported in the literature, are summarized in **Table 2**.

It has been shown that the presence of the aldehyde group on vanillin has an immediate effect on the thermal properties of the Bz monomer. Indeed, the presence of this group induces the formation of inter and intramolecular H bonds. The effect of inter and intramolecular H bonding is further highlighted by the high melting temperatures of vanillin-based di-Bz monomers. Indeed, the melting temperatures of the monomers reported in **Table 2** were not found to be lower than 145°C (**Table 2**, line 4), reaching even 229°C for the di-Bz with 4, 4′-diamino diphenyl sulfone (dds) as diamine (**Table 2**, line 2). These high melting temperatures, really close to the polymerization temperatures are impeding the processing of these monomers (melting and shaping). Nevertheless, vanillin-based Bz monomers led remarkably to cross-linked materials in spite of substituted phenolic *ortho* and *para* positions, impeding the polymerization. However, it was shown that the additional cross-linking reactions, due to the presence of an aldehyde group within a Bz monomer, occur mainly at the *ortho* position of the phenolic compound [39].

As the vanillin *ortho* position is blocked by a methoxy group, it can be assumed that the additional cross-linking of vanillin-based Bz monomers is mainly caused by the decarboxylation of the formyl group (**Figure 4**). Furthermore, some residual formyl groups, which did not undergo decarboxylation, are forming intermolecular H bonding, further increasing the PBz cross-linking density. Finally, the presence of the residual formyl groups is catalyzing the Bz monomers ROP [38, 39].

In consequence, due to the presence of the aldehyde functions and thus to the formation of inter- and intramolecular H bonds, vanillin-based materials were shown to display very high Tgs. Indeed, vanillin-based di-Bz monomers resulted in PBz networks with Tg values reaching 255°C (dds, **Table 2**, line 2). Nevertheless, di-vanillin Bz monomers are suffering from an evident drawback: their short processing windows. Indeed, melting temperatures of di-vanillin Bz monomers are too close to their polymerization temperatures, hindering their processing (molding and shaping) [38].

	Bz monomer	T_m (°C)	T_p (°C)	T_g (°C)	$T_{5\%}$ (°C)	Ref.
(1)	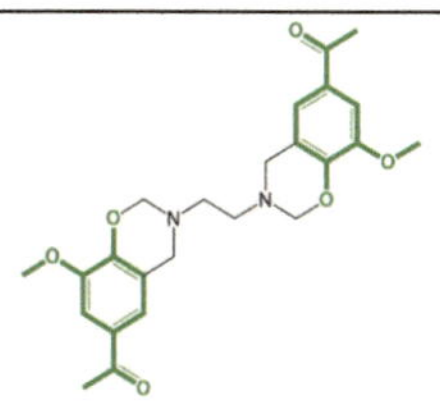 Vanillin/Ethylene diamine	205	213	–	283	[38]
(2)	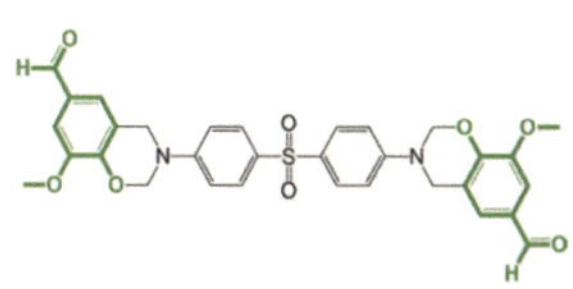 Vanillin/4,4′-diamino diphenyl sulfone (dds)	229	277	255	339	[38]
(3)	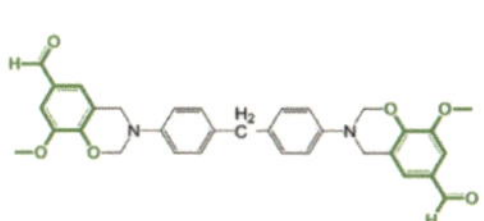 Vanillin/N,N diaminodiphenylmethane (ddm)	177	234	231	352	[38]
(4)	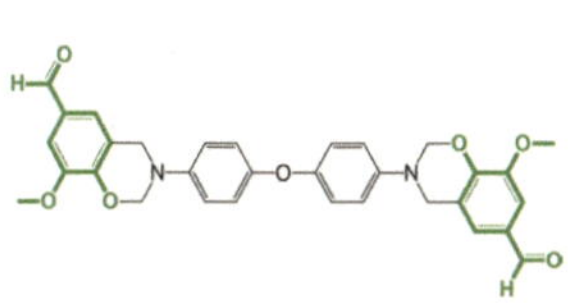 Vanillin/ (4-aminophenyl)ether (dde)	145	234	227	322	[38]
(5)	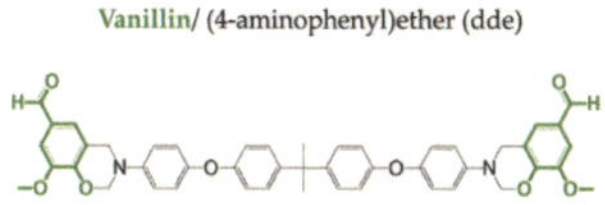 Vanillin/2,2-bis(4-(4-aminophenoxy)naphthalene (ndpa)	–	228	202	488	[38]

Table 2. Thermal properties of vanillin-based Bz monomers, synthetized from diamines, and T_g and $T_{5\%}$ of their corresponding PBz materials.

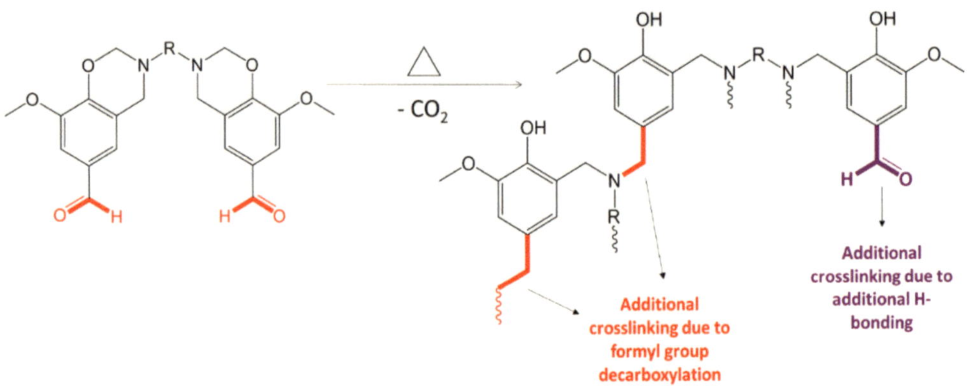

Figure 4. Cross-linking of vanillin-based di-Bz monomers [38].

3.3. From eugenol

Eugenol, through its availability and low cost, has also attracted attention for the synthesis of bio-based Bz monomers. Eugenol is the main component (72–90%wt) of the essential oil obtained from cloves. The chemical structure of eugenol is a disubstituted phenolic compound by a methoxy and allyl group at *ortho* and *para* positions, respectively. Consequently, due to blocked *ortho* and *para* positions, only the low reactive *meta* positions are available, impeding the formation of a cross-linked eugenol-based network (**Figure 5**).

Similar to di-Bz monomers from other naturally occurring phenolic compounds, di-Bz monomers from eugenol and various aromatic diamines were all displaying fairly high Tm > 90°C, hampering also the elaboration of eugenol-based PBz materials [40] (**Table 3**).

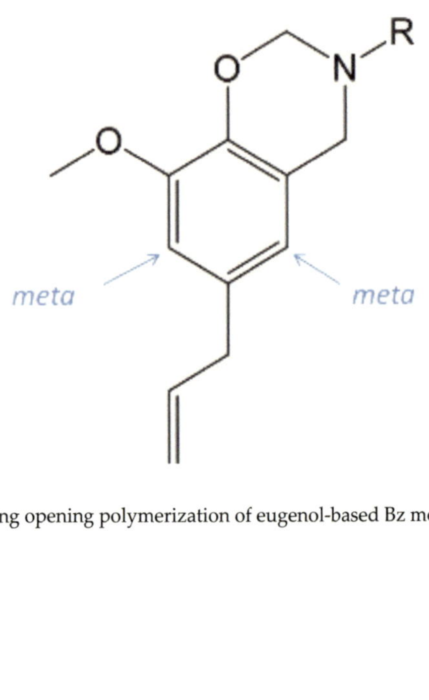

Figure 5. Available positions for ring opening polymerization of eugenol-based Bz monomer.

	Bz monomer	T_m (°C)	T_p (°C)	$T_{5\%}$ (°C)	Ref.
(1)	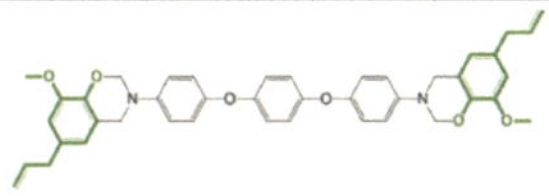 Eugenol/ (4-(4-aminophenoxy)phenyl)ether (44appe)	90	250	260	[40]
(2)	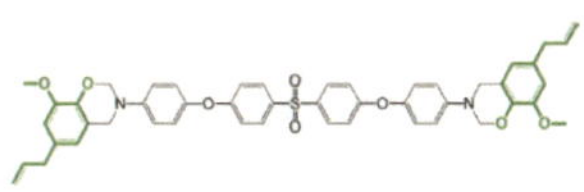 Eugenol/1,4(bis(4′-aminophenoxy)phenyl sulfone) (14apps)	100	250	275	[40]
(3)	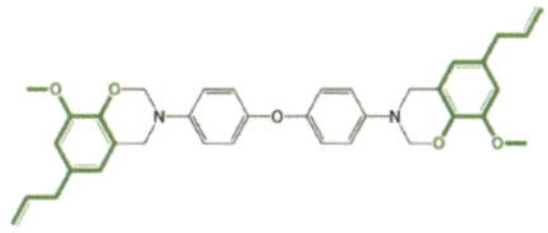 Eugenol/ (4-aminophenyl)ether (dde)	140	225	240	[40]
(4)	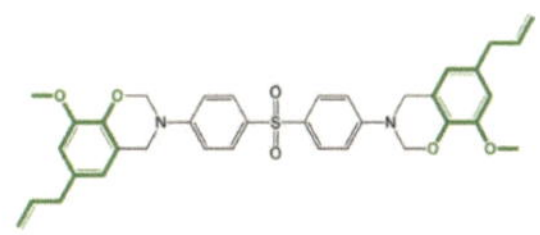 Eugenol/4,4′-diamino diphenyl sulfone (dds)	130	255	295	[40]
(5)	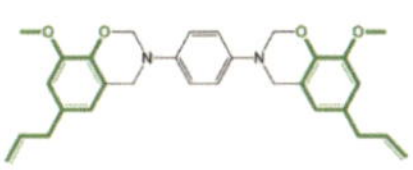 Eugenol/1,4-phenylene diamine	150	250	225	[41]

Table 3. Thermal properties of eugenol-based Bz monomers, synthetized from mono-functional amines or diamines, and $T_{5\%}$ of their corresponding PBz materials.

3.4. From other bio-phenols

Many other bio-phenols can be used for the design of benzoxazine monomers. Examples were reported with lignin-derivatives like coumaric acid, ferulic acid, phloretic acid [42], as well as coumarin [43], or urushiol [44, 45]. In most of the cases, only mono-benzoxazine has been prepared as the preparation of the monomers is hampered by substitution of the phenolic ring.

In conclusion, the use of naturally occurring phenolic compounds for the synthesis of Bz monomers allows the lowering of the carbon footprint and enables the introduction of some additional and/or functional groups within the Bz monomers. Nevertheless, most of the Bz monomers synthesized from bio-based phenolic compounds are suffering from several disadvantages, either for the synthesis of the Bz monomer, or the elaboration or the properties of the final materials.

An original and appropriate approach of adding different functionalities on Bz monomers consists of developing asymmetric di-Bz monomers from bio-phenols, using for example a diamine as bridging group between two phenolic derivatives exhibiting each a particular functionality. The different synthesis strategies developed to access asymmetric di-Bz monomers are presented in the following paragraphs.

4. Bio-based asymmetric benzoxazines

It is well known that additional functionalities on Bz monomers strongly influence the properties of the benzoxazine monomers and their corresponding thermoset materials [37, 40, 46]. The use of bio-based phenolic compounds naturally bearing various functionalities is thus considered as a clear asset to the synthesis of benzoxazine monomers with tailored properties. However, the presence of additional functional groups on Bz monomers may also impact their processability or polymerization. For instance, the position of the functional group on the phenolic compound, possible induced steric hindrance or short processing window due to high melting temperature of the monomer, may strongly hamper the processing and polymerization of the Bz monomers and thus affect the overall performance of the PBz materials. In consequence, drawbacks of some bio-based phenolic compounds could be offset by the assets of others to yield Bz monomers with large processing windows and corresponding high-performance PBz materials. To overcome such difficulties frequently encountered with Bz monomers synthetized with bio-based phenolics, some research groups reported the synthesis of asymmetric Bz monomers.

4.1. Eugenol/phenol asymmetric Bz

Among the available bio-based phenolic compounds, eugenol represents a high potential for the synthesis of Bz monomers. Nonetheless, the two main Bz polymerization sites of eugenol are hindered with methoxy and allyl groups in *ortho* and *para* positions, respectively. The elaboration of a highly cross-linked network is thus clearly hampered. Consequently, Dumas et al. proposed the preparation of a partially bio-based mixture of symmetric and asymmetric Bz monomers synthetized from phenol, eugenol, and 1,4-phenylenediamine, with (phenol: eugenol) ratio ranging from (1, 1) to (1.8, 0.2) [27, 28]. Three different chemical structures of Bz monomers, depicted in **Figure 6**, were found to constitute the mixture at the end of the reaction.

The synthesis yielded in two symmetric Bz monomers, a di-eugenol monomer and a di-phenol monomer, and one asymmetric Bz monomer showing both eugenol and phenolic moieties. The introduction of the synthesis of the asymmetric Bz monomer induced additional polymerization sites on the phenol moiety, increasing the cross-link density of the resulting material. Furthermore, the introduction of the eugenol part within the molecule allowed a solvent-free synthesis. Finally, the Tg increased from 120 to 220°C with the increasing (phenol: eugenol) ratio.

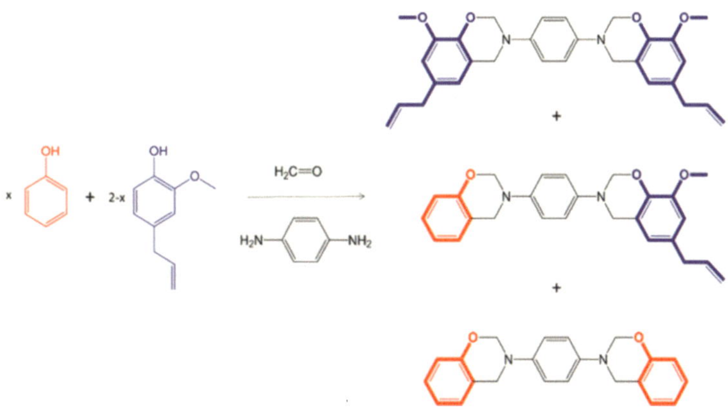

Figure 6. Synthesis of blends of di-eugenol, di-cardanol and asymmetric eugenol/cardanol di Bz monomers.

4.2. Vanillin/cardanol asymmetric Bz

Vanillin is another interesting and promising bio-based compound for the synthesis of Bz monomers. Due to the presence of the aldehyde function and thus due to the formation of inter- and intramolecular H bonds, vanillin-based Bz materials display high Tg. However, symmetric di-Bz monomer synthetized from vanillin and ethylenediamine (EDA) suffers from a major drawback: its melting temperature (Tm = 218°C) is too close to its polymerization temperature (Tp = 225°C) making its processing (molding and shaping) very challenging [47].

Cardanol, a bio-phenol bearing a C15 alkyl chain in *meta* position able to provide better processability and flexibility to the material through internal plasticization is relevant for the synthesis of usually brittle Bz materials. Cardanol di-Bz is of great interest due to low melting temperatures. The corresponding PBz material elaboration is nevertheless impeded by the steric hindrance of the alkyl side chain, resulting in long gelation time, low polymerization enthalpy value in DSC analysis, and a poorly cross-linked polymer material which creeps above Tg.

To tackle these highlighted issues, Puchot et al. proposed an efficient strategy to lower the melting temperature of vanillin-based Bz monomers in order to enable their processing [48]. Cardanol was shown to provide flexibility to the resulting materials through plasticizing effect of the C15 alkyl chain and to result in a Bz monomer with low melting temperature [33, 49]. More importantly, the low reactivity of cardanol offers the possibility to control the synthesis of a mono-cardanol Bz with a free primary amine moiety that can be further coupled with another bio-based phenol like vanillin. The authors aimed at synthetizing an asymmetric bio-phenol-based di-Bz monomer by combining cardanol and vanillin via a two-step synthesis pathway with ethylene diamine including a controlled mono-substitution of a cardanol-based Bz, followed by a coupling with vanillin (**Figure 7**).

Such combination resulted in a Bz monomer with low melting temperature (Tm = 101°C), with a few impact on the activation temperature (Tp = 227°C) increasing the processing window of these bio-based thermosets (ΔTproc = 126°C, compared to ΔTproc = 7°C for di-vanillin Bz).

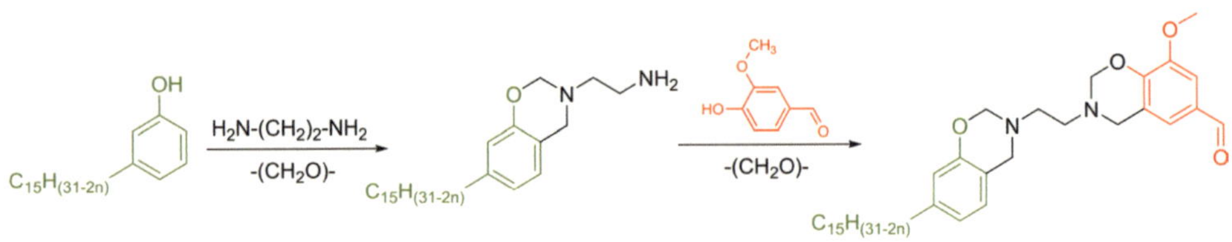

Figure 7. Representative scheme of the synthesis of vanillin/cardanol asymmetric Bz.

Furthermore, a second drawback of the di-vanillin Bz monomers was their thermal degradation occurring simultaneously to their melting and even before their polymerization ($T_{5\%}$ = 203°C). In the case of the asymmetric Vani-Card Bz monomer, no problem of thermal degradation of the monomer is further encounter due to the presence of the cardanol moiety. Indeed, the thermal degradation of the Vani-Card monomer occurred at higher temperature than melting and polymerization ($T_{5\%}$ = 280°C).

The asymmetric Bz monomer enabled thus, through a processing point of view, an easy elaboration of a cardanol- and vanillin-based PBz material. Furthermore, the synthesis of the asymmetric Vani-Card monomer was also shown to strongly influence the reactivity of the monomer. Indeed, the di-cardanol Bz monomer displayed a long gelation time (Tgel = [38, 40] min at 190°C) in comparison to a classical di-phenol Bz monomer (Tgel = [5, 6] min at 190°C). In the case of Vani-Card Bz, an interesting moderate gelation time of 8–9 min was reported.

Moreover, the characterizations of the asymmetric Vani-Card PBz material highlighted the obtaining of a thermoset material with good thermo-mechanical properties. Indeed, a Tg of 129°C and a storage modulus in the glassy state of 700 MPa were reported, indicating that Vani-Card PBz material displayed very comparable thermal and mechanical properties to other known polybenzoxazines particularly those partially or fully bio-based.

Finally, the presence of the cardanol moiety was shown to strongly influence the thermal stability of the PBz material. Indeed, a high thermal degradation onset temperature was recorded for Vani-Card PBz material ($T_{5\%}$ = 373°C) compared to classical di-phenol PBz material ($T_{5\%}$ = 273°C).

This versatile approach is a powerful tool to access a wide range of bio-based asymmetric benzoxazines given the wide number of bio-phenols available. It paves the way toward easily processable high performance bio-based benzoxazines. This strategy could, for example, be applied to further improve the processing of bio-based di-Bz monomers.

5. End-chain functionalization polymers

Beyond the elaboration of symmetric and asymmetric benzoxazine monomers, a recent challenge is to add benzoxazine functionalities to polymers, benefiting then from the versatile design of Bz monomers. The resultant materials possess properties specific both to benzoxazine, such as a high cross-linking density and high temperature stability, and to the backbone polymer, like processability and flexibility. End-chain, [50] main-chain [51–53], as well as side-chain [54] benzoxazine polymers have been reported. Therefore, Bz monomers have been anchored onto a wide range of (bio)polymers like polybutadiene [55], polystyrene [56], polypropylene oxide [57], polyvinylchloride [58], cellulose [54], poly(ε-caprolactone) [59], and even onto macromonomers like lignin [60]. In almost all cases, harmful phenol has been used as a reactant to form the oxazine ring.

Naturally occurring phenolic compounds can also be used to functionalize the end-chain of existing polymers toward benzoxazine ring formation. Phloretic acid (PA) is an interesting bio-based compound for this purpose due to the presence of a propionic acid on the *para* position suitable for esterification [61]. Furthermore, *ortho* positions of PA are not substituted, an ideal asset for the design of benzoxazine monomers and for their polymerization. Trejo-Machin et al. proposed a sustainable strategy to functionalize hydroxyls (–OH)-bearing molecules. Ethylene glycol, and two polyethylene glycol with molar masses about 400 g·mol^{-1} (PEG400) and 2000 g·mol^{-1} (PEG2000), were reacted by esterification with PA. In the second step, these phenolic groups were reacted with furfurylamine to form benzoxazine end-capped molecules. The two-step synthesis of end-chain Bz monomers did not require neither solvent nor purification and led to a set of materials almost 100% bio-based (**Figure 8**) [62].

Thermal behavior and stability of Bz end-chain monomers were characterized by DSC and TGA, demonstrating the elaboration of thermoset materials. Moreover, DSC analyses showed on each thermogram an exothermic peak assigned to the thermal-activated ring-opening of the Bz monomers, with an onset ranging from 165 to 200°C. Rheological analyses were also carried out to evaluate the polymerization reactivity of the Bz monomers. The gelation time at 180°C was found to be 2, 3, and 28 min for benzoxazine prepared with ethylene glycol, PEG400, and PEG2000, respectively. The mechanical relaxation temperatures were decreasing from 112 to 27°C and −45°C, with increasing backbone chain length of the starting molecules, respectively. Finally, the thermal stability of the three PBz materials was studied by TGA analyses. Each of the synthetized PBz material exhibited much higher thermal stability as well as marked improvement in char yield over the origin polymers.

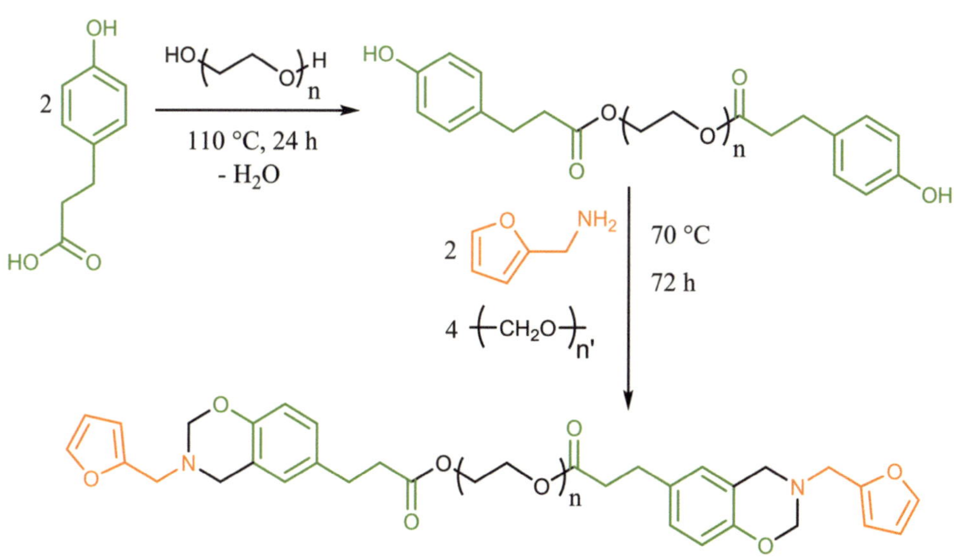

Figure 8. Two-step synthesis of end-chain benzoxazine monomers; n = 1, 9, and 45 and n′ = 8–100.

In conclusion, this innovative approach proved the suitability of phloretic acid to act as a renewable building block to functionalize hydroxyl-bearing molecules and polymers toward the synthesis of end-chain benzoxazine polymers.

6. Conclusions

The use of naturally occurring phenolic compounds for the synthesis of Bz monomers allows the lowering of the carbon footprint and enables the introduction of some additional and/or functional groups within the Bz monomers which can highly benefit the industrial applications of Bz where additional functional groups may be needed in order to enhance their properties. Alike petroleum-based Bz monomers, the nature of the diamine bridging group strongly affects the bio-based Bz monomers properties and their polymers. Short aliphatic diamine bridging groups, such as ethylene diamine, should as well be preferred for the synthesis of bio-based Bz monomers. Indeed, the use of short aliphatic diamines promotes moderate processing windows, while resulting in high cross-linked PBz materials with good thermal and mechanical properties.

Nevertheless, most of the Bz monomers synthetized from bio-based phenolic compounds are suffering from several disadvantages, either for the synthesis of the Bz monomer or for the elaboration and/or for the properties of the final materials. Naturally occurring phenolic compounds such as lignin-derivatives or coumarin compounds lead to highly cross-linked PBz materials with good thermo-mechanical properties, whereas their processing is hindered by narrow processing windows. On the contrary, bio-based Bz monomers synthetized from phenolic compounds like cardanol, urushiol, or eugenol display large processing windows but low cross-linking density and low Tg. As the overall performance of PBz is strongly affected by oxazine ring substitutions, drawbacks of some bio-based phenolic compounds could be offset by the assets of the others to yield Bz monomers with large processing windows and corresponding high-performance PBz materials.

Conflict of interest

There is no conflict of interest to declare.

Author details

Acerina Trejo-Machin[1,2], Laura Puchot[1] and Pierre Verge[1*]

*Address all correspondence to: pierre.verge@list.lu

1 Luxembourg Institute of Science and Technology, Esch-sur-Alzette, Luxembourg

2 University of Luxembourg, Esch-sur-Alzette, Luxembourg

References

[1] Ratna D. Handbook of Thermoset Resins. Shawbury, UK: ISmithers; 2009

[2] Knop A, Pilato LA. Phenolic Resins: Chemistry, Applications and Performance. Berlin: Springer Science & Business Media; 2013

[3] Gardziella A, Pilato LA, Knop A. Phenolic Resins: Chemistry, Applications, Standardization, Safety and Ecology. Berlin: Springer Science & Business Media; 2013

[4] Ishida H. Chapter 1—Overview and Historical Background of Polybenzoxazine Research. Amsterdam: Elsevier; 2011. pp. 3-81

[5] Ning X, Ishida H. Journal of Polymer Science Part B: Polymer Physics. 1994;**32**:921-927

[6] Ishida H, Low HY. Macromolecules. 1997;**30**:1099-1106

[7] Kanchanasopa M, Yanumet N, Hemvichian K, Ishida H. Polymers and Polymer Composites. 2001;**9**:367-376

[8] Shen SB, Ishida H. Polymer Composites. 1996;**17**:710-719

[9] Ishida H, Allen DJ. Journal of Polymer Science Part B: Polymer Physics. 1996;**34**:1019-1030

[10] Verge P, Puchot L, Vancaeyzeele C, Vidal F, Habibi Y. In: Froimowicz P, editor. Chapter 7—Symmetric Versus Asymmetric di-Bz Monomer Design: Structure-to-Properties Relationship A2—Ishida, Hatsuo. Amsterdam: Elsevier; 2017. pp. 89-107

[11] Zhou C, Lu X, Xin Z, Liu J. Corrosion Science. 2013;**70**:145-151

[12] Zhou C, Lu X, Xin Z, Liu J, Zhang Y. Progress in Organic Coatings. 2013;**76**:1178-1183

[13] Zhou C, Lu X, Xin Z, Liu J, Zhang Y. Corrosion Science. 2014;**80**:269-275

[14] Lin S-C, Wu C-S, Yeh J-M, Liu Y-L. Polymer Chemistry. 2014;**5**:4235-4244

[15] Escobar J, Poorteman M, Dumas L, Bonnaud L, Dubois P, Olivier M-G. Progress in Organic Coatings. 2015;**79**:53-61

[16] Lligadas G, Tuzun A, Ronda JC, Galia M, Cadiz V. Polymer Chemistry. 2014;**5**:6636-6644

[17] Bălănucă B, Raicopol M, Maljusch A, Garea S, Hanganu A, Schuhmann W, et al. ChemPlusChem. 2015;**80**:1170-1177

[18] Ishida H, Agag T. Handbook of Benzoxazine Resins. Elsevier; 2011

[19] Ishida H, Froimowicz P. Elsevier; 2017. pp. 528-531

[20] Ning X, Ishida H. Journal of Polymer Science Part A: Polymer Chemistry. 1994;**32**:1121-1129

[21] Wang C, Sun J, Liu X, Sudo A, Endo T. Green Chemistry. 2012;**14**:2799-2806

[22] Reiss G, Schwob J, Guth G, Roche M, Lande B. In: Culbertson BM, McGrath JE, editors. Advances in Polymer Synthesis. New York: Plenum; 1985

[23] Liu C, Chen QY. In: Froimowicz P, editor. Chapter 2—Catalytic Accelerated Polymerization of Benzoxazines and Their Mechanistic Considerations A2–Ishida, Hatsuo. Amsterdam: Elsevier; 2017. pp. 9-21

[24] Wang YX, Ishida H. Polymer. 1999;**40**:4563-4570

[25] Ishida H, Rodriguez Y. Journal of Applied Polymer Science. 1995;**58**:1751-1760

[26] Sini NK, Bijwe J, Varma IK. Polymer Degradation and Stability. 2014;**109**:270-277

[27] Dumas L, Bonnaud L, Olivier M, Poorteman M, Dubois P. Journal of Materials Chemistry A. 2015;**3**:6012-6018

[28] Thirukumaran P, Shakila A, Muthusamy S. RSC Advances. 2014;**4**:7959-7966

[29] Lochab B, Varma IK, Bijwe J. Journal of Thermal Analysis and Calorimetry. 2012;**107**: 661-668

[30] Lochab B, Varma IK, Bijwe J. Journal of Thermal Analysis and Calorimetry. 2010;**102**: 769-774

[31] Agag T, An SY, Ishida H. Journal of Applied Polymer Science. 2013;**127**:2710-2714

[32] Lochab B, Varma IK, Bijwe J. Journal of Thermal Analysis and Calorimetry. 2010;**102**: 769-774

[33] Lochab B, Varma IK, Bijwe J. Journal of Thermal Analysis and Calorimetry. 2012;**107**: 661-668

[34] Campaner P, D'Amico D, Longo L, Stifani C, Tarzia A, Tiburzio S. In: Agag T, editor. Chapter 19—Study of a Cardanol-Based Benzoxazine as Reactive Diluent and Toughening Agent of Conventional Benzoxazines A2—Ishida, Hatsuo. Amsterdam: Elsevier; 2011. pp. 365-375

[35] Araújo JDP, Grande CA, Rodrigues AE. Chemical Engineering Research and Design. 2010;**88**:1024-1032

[36] Brazinha C, Barbosa DS, Crespo JG. Green Chemistry. 2011;**13**:2197-2203

[37] Van A, Chiou K, Ishida H. Polymer. 2014;**55**:1443-1451

[38] Sini NK, Bijwe J, Varma IK. Polymer Degradation and Stability. 2014;**109**:270-277

[39] Ran Q-C, Gu Y. Journal of Polymer Science Part A: Polymer Chemistry. 2011;**49**:1671-1677

[40] Thirukumaran P, Shakila A, Muthusamy S. RSC Advances. 2014;**4**:7959-7966

[41] Dumas L, Bonnaud L, Olivier M, Poorteman M, Dubois P. European Polymer Journal. 2015;**67**:494-502

[42] Comí M, Lligadas G, Ronda JC, Galià M, Cádiz V. Journal of Polymer Science Part A: Polymer Chemistry. 2013;**51**:4894-4903

[43] Arza CR, Froimowicz P, Ishida H. RSC Advances. 2015;**5**:97855-97861

[44] Xu H, Lu Z, Zhang G. RSC Advances. 2012;**2**:2768-2772

[45] Xu H, Zhang W, Lu Z, Zhang G. RSC Advances. 2013;**3**:3677-3682

[46] Santhosh Kumar KS, Reghunadhan Nair CP, Radhakrishnan TS, Ninan KN. European Polymer Journal. 2007;**43**:2504-2514

[47] Sini NK, Bijwe J, Varma IK. Journal of Polymer Science Part A: Polymer Chemistry. 2014;**52**:7-11

[48] Puchot L, Verge P, Fouquet T, Vancaeyzeele C, Vidal F, Habibi Y. Green Chemistry. 2016;**18**:3346-3353

[49] Rao BS, Palanisamy A. European Polymer Journal. 2013;**49**:2365-2376

[50] Chernykh A, Agag T, Ishida H. Polymer. 2009;**50**:3153-3157

[51] Shibayama Y, Kawauchi T, Takeichi T. High Performance Polymers. 2014;**26**:60-68

[52] Takeichi T, Kano T, Agag T. Polymer. 2005;**46**:12172-12180

[53] Kiskan B, Yagci Y, Ishida H. Journal of Polymer Science Part A: Polymer Chemistry. 2008;**46**:414-420

[54] Agag T, Vietmeier K, Chernykh A, Ishida H. Journal of Applied Polymer Science. 2012;**125**:1346-1351

[55] Kukut M, Kiskan B, Yagci Y. Designed Monomers and Polymers. 2009;**12**:167-176

[56] Ergin M, Kiskan B, Gacal B, Yagci Y. Macromolecules. 2007;**40**:4724-4727

[57] Yildirim A, Kiskan B, Demirel AL, Yagci Y. European Polymer Journal. 2006;**42**:3006-3014

[58] Kiskan B, Demiray G, Yagci Y. Journal of Polymer Science Part A: Polymer Chemistry. 2008;**46**:3512-3518

[59] Kiskan B, Yagci Y. Polymer. 2005;**46**:11690-11697

[60] Abarro GJ, Podschun J, Diaz LJ, Ohashi S, Saake B, Lehnen R, et al. RSC Advances. 2016;**6**:107689-107698

[61] Comí M, Lligadas G, Ronda JC, Galià M, Cádiz V. Journal of Polymer Science Part A: Polymer Chemistry. 2013;**51**:4894-4903

[62] Trejo-Machin A, Verge P, Puchot L, Quintana R. Green Chemistry. 2017;**19**:5065-5073

Chapter 5

UV and Thermal Cure Epoxy Adhesives

Chunfu Chen, Bin Li, Chao Wang, Shuichi Iwasaki,
Masao Kanari and Daoqiang Lu

Additional information is available at the end of the chapter

http://dx.doi.org/10.5772/intechopen.82168

Abstract

Typical commercial UV and thermal cure epoxy adhesives have been reviewed and compared. UV cure cationic epoxy adhesives are primarily composed of cycloaliphatic epoxy resin and cationic photoinitiator. UV cationic epoxy adhesives have no surface cure issue and possess low cure shrinkage and good adhesion performance but need post-thermal cure to achieve full adhesion performance in use. Hybrid UV acrylate and thermal cure epoxy adhesives are primarily composed of acrylate monomer, free radical photoinitiator, epoxy resin and curing agent. The hybrid epoxy adhesives combine fast UV curability of acrylate composition and high adhesion performance of thermal cure epoxy composition. A new type initiator free hybrid one-component UV and thermal cure adhesive has been also introduced. It is mainly composed of maleimide compound, acrylic monomer, partially acrylated epoxy resin, epoxy resin and latent curing agent. Its UV cure and thermal cure behaviour have been studied by FT-IR spectroscopy measurement.

Keywords: UV cure, thermal cure, cationic, free radical, acrylate, epoxy adhesive

1. Introduction

Epoxy adhesives are widely used in structural bonding applications ranging from general industry, semiconductor packaging, electronics assembly and automobile production to aerospace market because of their strong chemical structure and good adhesion to various substrates [1–9]. Epoxy adhesives are primarily composed of epoxy resin and curing agent. **Figure 1** illustrates chemical structure and key features of various functional groups for bisphenol A diglycidyl ether, the most standard epoxy resin used in epoxy adhesives. Epoxide possesses high reactivity. It can react with amines, thiols, anhydrates or phenols almost equivalently via polyaddition mechanism at suitable certain conditions to become strong cross-linked thermoset

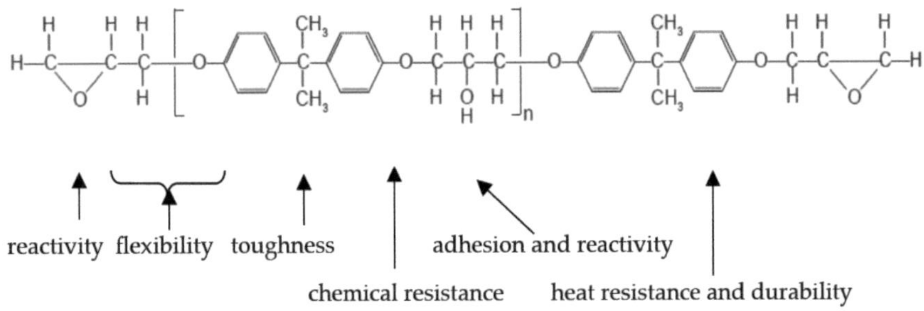

Figure 1. Structure and key features of bispheol A diglycidyl ether.

resins. As shown in **Scheme 1** [10], epoxide reacts almost equivalently with active hydrogen in amine curing agent via polyaddition mechanism. Epoxide can also polymerize via either anionic or cationic polymerization mechanism. As shown in **Scheme 2** [11], epoxide can polymerize via anionic polymerization mechanism initiated by anionic ion resulted from reaction of imidazole compound and epoxide. Epoxy adhesives can be cured at different temperature conditions based mainly on the curing agent type used but will normally need relatively long cure time ranging from half hour to a few days. Aliphatic amine-based epoxy adhesives, the most commonly used type, start to cure at room temperature. Thiol-based epoxy adhesives, the fastest cure type, start to cure even at low refrigerator temperature. Anhydrate-, phenol-, aromatic amine- or catalyst-based epoxy adhesives will normally need elevated temperature to achieve full cure. Generally speaking, epoxy adhesives designed to cure at elevated temperature which are commonly called as thermal cure epoxy adhesives have higher degree of cross-linking structure and glass transition temperature and thus show better performance than epoxy adhesives designed for cure at room temperature. Thermal cure epoxy adhesives can be also formulated as one-component type by the use of latent curing agents for easy handling. One-component thermal cure epoxy adhesives have been increasingly used in various applications such as semiconductor packaging, electronics assembly and automobile production where high production efficiency and high adhesion performance are required. Recently, UV and thermal cure epoxy adhesives have been

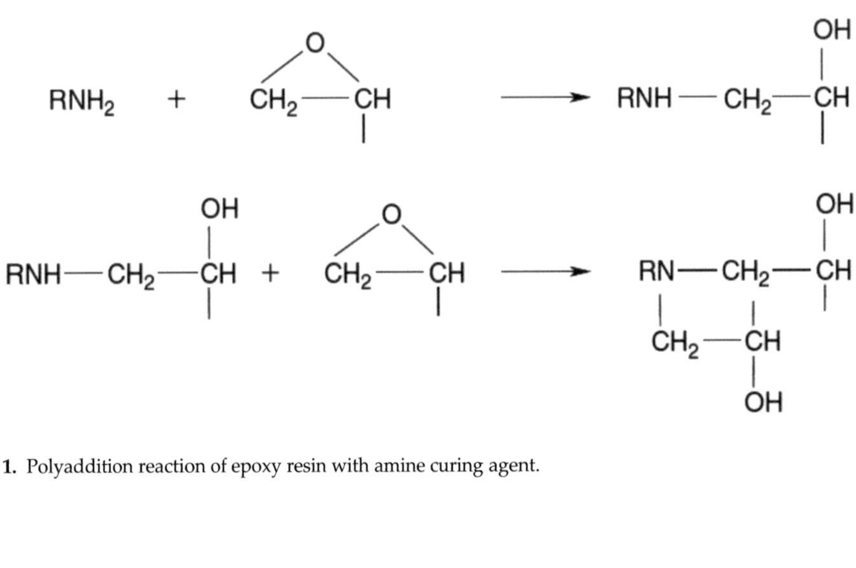

Scheme 1. Polyaddition reaction of epoxy resin with amine curing agent.

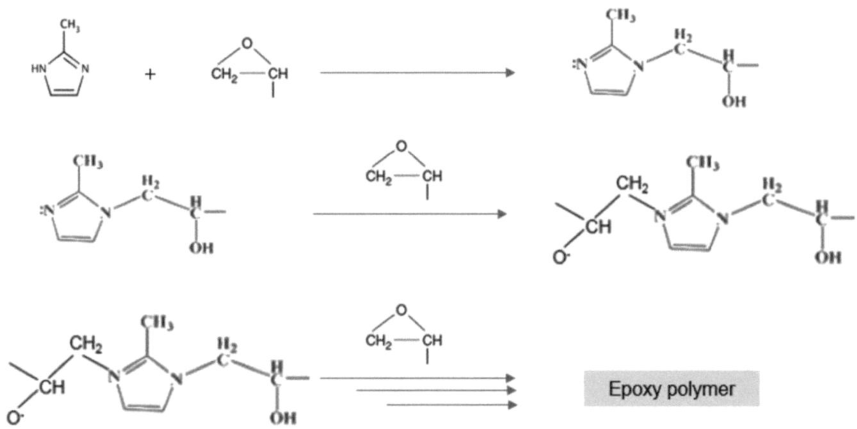

Scheme 2. Anionic polymerization of epoxy resin via imidazole catalyst.

developed and commercialized to meet further higher production efficiency of required applications such as precise optical sensor packaging and display assembly [12–15].

There are mainly two types of commercial UV and thermal cure epoxy adhesives: UV cure cationic epoxy adhesives and hybrid UV acrylate and thermal cure epoxy adhesives. UV cationic cure epoxy adhesives are primarily composed of cycloaliphatic epoxy resin and cationic photoinitiator. UV cationic epoxy adhesives have no surface cure issue and possess low cure shrinkage and good adhesion performance but need post-thermal cure to achieve full cure. Hybrid UV acrylate and thermal cure epoxy adhesives are primarily composed of acrylate monomer, free radical photoinitiator, epoxy resin and curing agent. The hybrid epoxy adhesives combine fast UV curability of acrylate composition and high adhesion performance of thermal cure epoxy composition.

2. UV cationic epoxy adhesives

UV cationic epoxy adhesives are primarily composed of epoxy resin and cationic photoinitiator [16–20]. Cycloaliphatic-type epoxy resins are usually selected for UV cationic epoxy adhesives because of faster cationic polymerization rate than that of normal bisphenol A diglycidyl ether-type epoxy resin. Chemical structure of typical commercially available epoxy resins suitable for cationic epoxy adhesives is shown in **Figure 2**. Cationic photoinitiator is the key raw material to formulate UV cationic epoxy adhesives. There are mainly two types of cationic photoinitiators: Bronsted acid and Lewis acid generator. Sulfonium and iodonium salts that can generate Bronsted acid are most commonly used as cationic photoinitiator. **Figure 3** shows chemical structure of typical commercially available cationic photoinitiators.

As illustrated in **Scheme 3** [21], photoinitiator in UV epoxy adhesives absorbs UV energy to generate strong acid that will react with epoxy to produce cationic which can initiate homo-polymerization of epoxy resin. UV cationic epoxy adhesives will need some longer cure

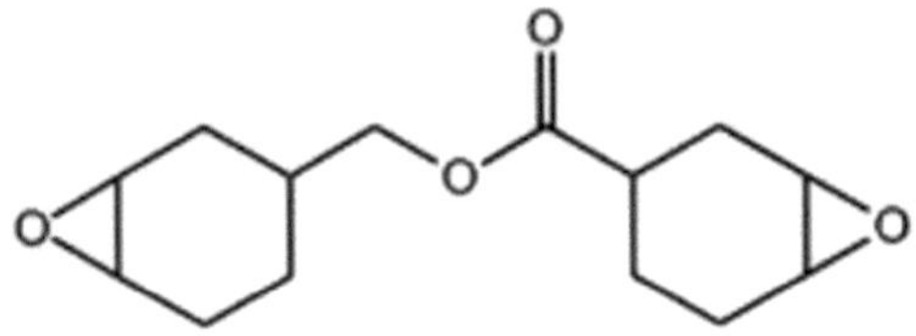

3, 4-epoxycyclohexylmethyl-3,4-epoxycyclohexane carboxylate

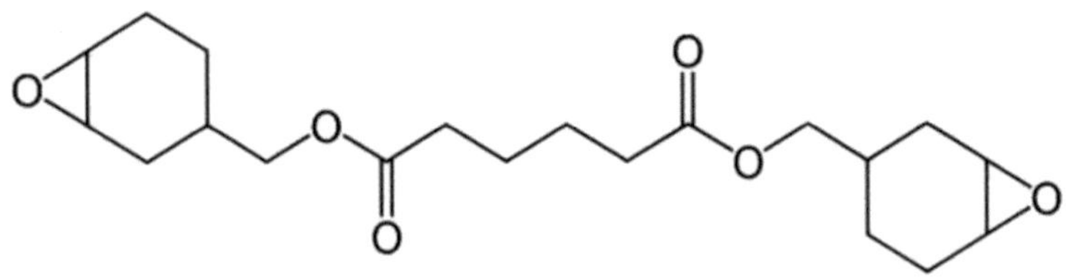

bis(3,4-epoxycyclohexylmethyl) adipate

Figure 2. Common commercially available epoxy resins for cationic epoxy adhesives.

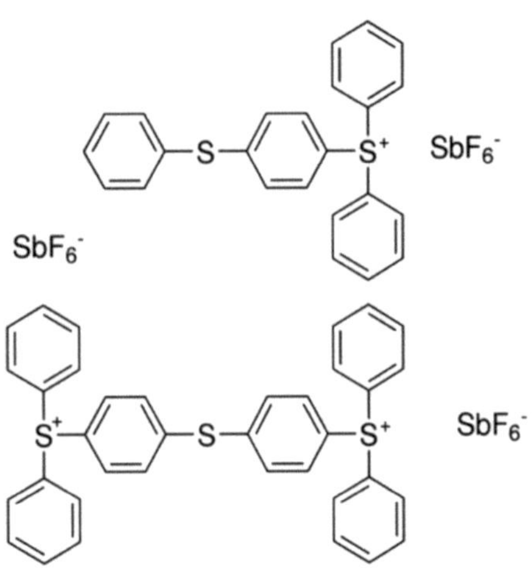

mixed triarylsulfonium hexafluoroantimonate salts

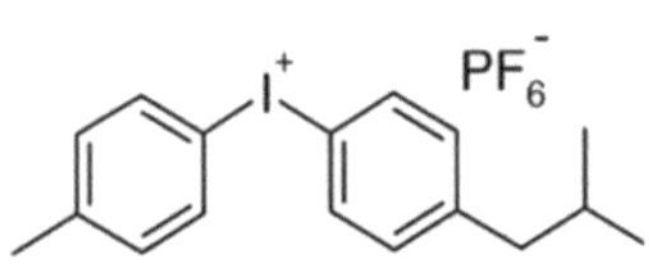

(4-Methylphenyl) [4-(2-methylpropyl) phenyl] iodonium hexafluorophosphate

Figure 3. Chemical structure of common UV cationic photoinitiator.

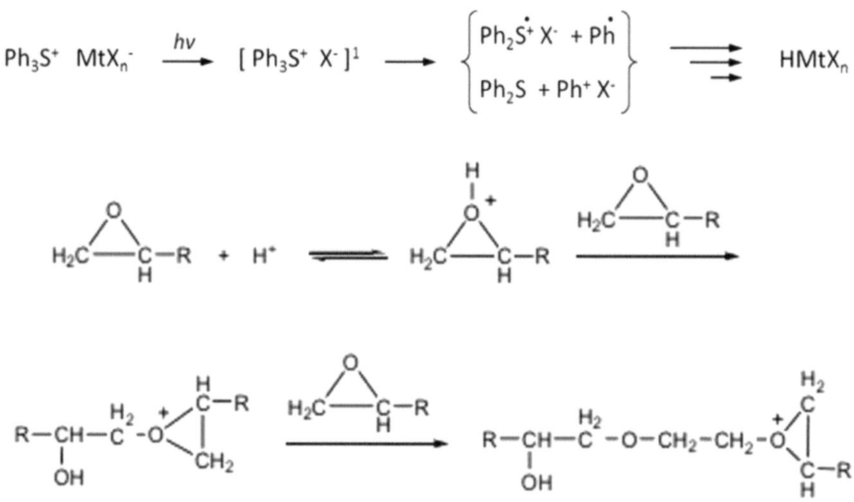

Scheme 3. UV cationic polymerization of epoxy adhesives.

time compared to UV cure acrylate-based adhesive. In actual use, a post-thermal cure of UV cationic epoxy adhesives after the UV radiation is commonly used for full cure to assure satisfactory adhesion performance. Compared to common acrylate-based UV adhesives, UV cationic epoxy adhesives have much lower cure shrinkage because of the epoxy structure and have no surface cure issue that is resulted from oxygen inhibition to free radical polymerization since they cure via cationic polymerization. By contrast, UV cationic epoxy adhesives are not suitable for alkali-type substrates which stop cationic polymerization.

UV cationic epoxy adhesives have been commercialized and used in optical parts bonding, sensor packaging and display panel assembly applications [22–26]. The authors have found that adhesion reliability performance of UV cationic epoxy adhesives can be much improved by the combination use of cationic photoinitiator with thermal cationic initiator [27].

3. Hybrid UV acrylate and thermal cure epoxy adhesives

Most widely used UV cure adhesives are acrylate-based compositions [28–32]. Acrylate-based UV cure adhesives are primarily composed of acrylate monomer, acrylate oligomer and photoinitiator. As shown in **Scheme 4** [33], the photoinitiator formulated in an acrylate-based adhesive absorbs light energy via UV radiation to generate free radical which can rapidly initiate polymerization of acrylate compositions. Acrylate-based UV cure adhesives can be cured within seconds. Limitations of UV cure acrylate-based adhesives are the surface cure issue, shadow cure problem, high cure shrinkage and poor humidity reliability. Surface cure issue is resulted from oxygen inhibition to free radical polymerization of acrylate. Shadow cure problem always occurs at the area where light cannot approach. Relatively high cure shrinkage and poor humidity reliability are caused from acrylate chemical structure.

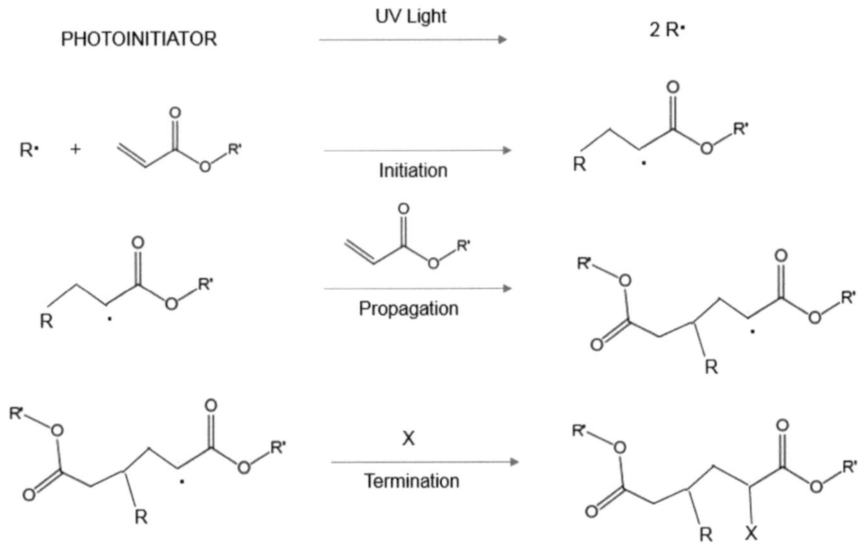

Scheme 4. UV cure mechanism of free radical polymerization of acrylate adhesives.

By the combination of UV acrylate composition with thermal cure epoxy composition, UV and thermal cure hybrid epoxy adhesives have been developed and commercialized for over two decades [34–38]. Acrylate monomer, epoxy resin, photoinitiator and epoxy curing agent are at least contained in the UV and thermal cure hybrid adhesives. These hybrid adhesives combine advantages from both UV acrylate proportion and thermal cure epoxy part. Adhesion reliability

Adhesive type	UV acrylate	UV cationic epoxy	Hybrid thermal cure epoxy
Key compositions	Acrylate	Epoxy resin	Acrylate
	Photoinitiator	Cationic photoinitiator	Photoinitiator
			Epoxy resin
			Curing agent
Polymerization			
UV cure	Radical	Cationic	Radical
Thermal cure	NA	Cationic	Polyaddition, anionic
Oxygen inhibition	Yes	No	Partially
Alkali inhibition	No	Yes	No
UV curability	High	Medium	High
Post-thermal cure	No need	Preferred	Need
Shadow cure	No	Partially	Yes
Cure shrinkage	High	Low	Low
Adhesion	Moderate	Good	Good

Table 1. Comparison of UV acrylate, cationic epoxy and hybrid thermal cure epoxy.

performance could be much improved by the introduction of the epoxy composition compared to the normal acrylate composition. In the meantime, production efficiency could be much improved by shortening the fixture time to seconds via UV cure compared to at least dozens of minutes needed for thermal cure epoxy adhesives. Surface cure issue, shadow cure issue and high cure shrinkage of acrylate-based UV adhesives could also be improved to certain degree because of lower contents of free radical curable acrylate compositions. In some cases, a thermal initiator such as peroxide is also formulated in the hybrid adhesive to assure curing remained acrylate compositions after the UV radiation or those at shadow area. Advantages and limitations of UV cationic epoxy adhesives, hybrid UV acrylate and thermal cure epoxy adhesives are compared with those of UV acrylate adhesives in **Table 1**.

4. Initiator free hybrid epoxy adhesives

Photoinitiator is the key material to formulate UV cure compositions. In actual cure process, however, several small molecules are usually generated as byproducts. Additionally, photoinitiator itself will not be consumed completely in actual use at most cure conditions and will remain in the cured materials as just contaminants. As shown in **Scheme 5**, for the use of benzyl dimethyl ketal (BDMK) as photoinitiator, for example, Sitmann et al. [39] described that there are at least three small molecules generated during its UV light decomposition. These small molecular byproducts, together with the remained photoinitiator, cannot be chemically bonded to the cured adhesive. For sensitive high precise substrate bonding applications such as fine semiconductor packaging or display assembly, there are big concerns on contaminants from low molecule chemicals such as these byproducts, remained photoinitiator during UV curing process on sensitive semiconductor substrate or display materials. In addition, the remained photoinitiator

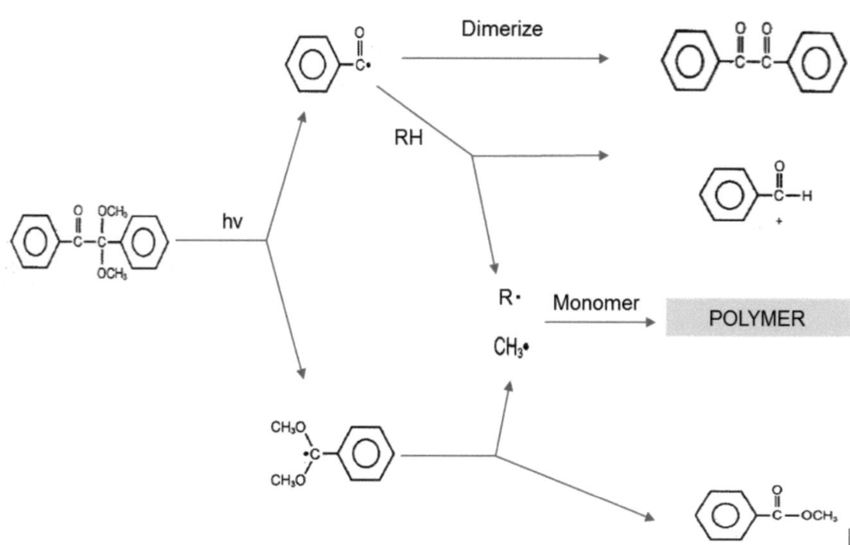

Scheme 5. Photo-reaction mechanism of BMDK.

may initiate or accelerate chemical reaction of cured adhesive materials during the actual use and potentially damage its adhesion reliability performance. Initiator free UV cure adhesive will not have these concerns.

Recently, the authors invented and reported a new type high-performance UV and thermal curable hybrid epoxy adhesive that is completely an initiator free composition but still possesses good UV curability and satisfactory thermal curability, suitable for use in high-end display assembly applications [40–42].

Maleimide compounds have been studied for years in photoinitiator-free UV curing systems [43–46]. As illustrated in **Scheme 6**, maleimide compound can adsorb light energy and generate small amount of free radical. In the meantime, maleimide itself is a good monomer for free radical polymerization. Compared to normal photoinitiator acrylate cases, however, its UV cure efficiency is much lower.

The new type hybrid epoxy resin adhesive is mainly composed of a liquid bismaleimide compound, partially acrylated bisphenol A epoxy resin, acrylic monomer, epoxy resin and latent curing agent. Chemical structure of typical reactive materials used is shown in **Figure 4**. The new type adhesive does not contain any conventional initiator, either photoinitiator or thermal initiator such as peroxide compound. It is a complete initiator free hybrid epoxy adhesive. Its UV fixture time was 5 s at 100 mW/cm^2 with high-pressure mercury lamp used. Good adhesion on glass substrate has been also confirmed.

FT-IR was performed to measure and analyze quantitatively cure behaviour of the adhesive sample [46–48]. The spectrum of adhesive samples cured at UV cure and UV + thermal cure conditions as well as non-cure samples was measured by the use of Varian 610-IR Fourier

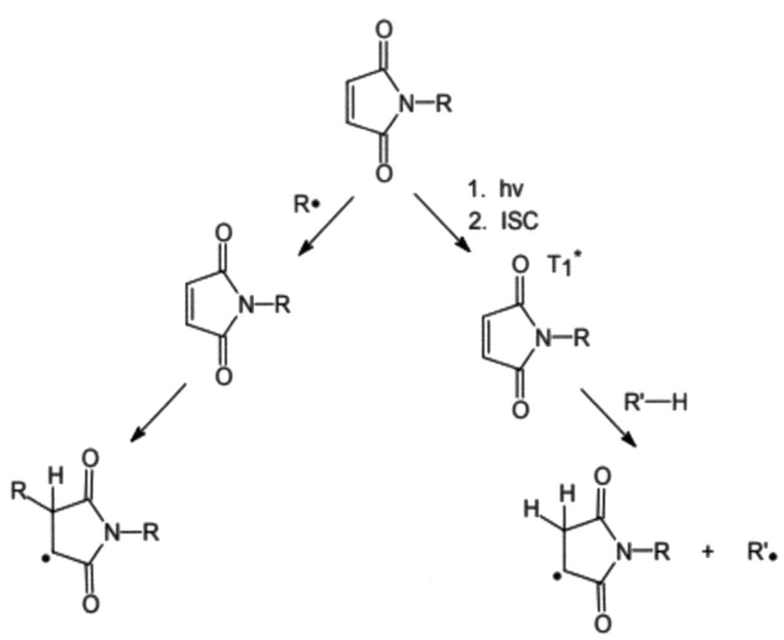

Scheme 6. Radical generation of maleimides via UV radiation.

transform infrared (FT-IR) spectroscopy. **Figure 5** IR spectrum of adhesive sample cured at UV only condition was shown compared to non-cure sample. IR spectrum of adhesive samples cured at UV + thermal cure condition was shown compared to non-cure sample in **Figure 6**. The conversion rate was further calculated from the decrease of the 1405 cm^{-1} absorption peak area attributed to acrylic double bond, the 690 cm^{-1} peak area attributed to maleimide double bond and the 915 cm^{-1} peak area attributed to epoxy group. As summarized in **Table 2**, a conversion rate of 62% of acrylic and 95% of maleimide double bonds had been achieved at this UV cure condition. This result confirmed that most part of acrylic and almost all maleimide double bonds had been cured during this UV cure condition. As expected, epoxy group cured only at thermal cure condition.

Very interestingly, it was found, as shown in **Table 2**, that remained uncured acrylic double bonds at UV cure process continued to react, and the conversion rate increased eventually to 100% at post-thermal cure condition. In the meantime, conversion rate of acrylic double achieved 67 and 95% at thermal cure only condition, respectively. As described previously,

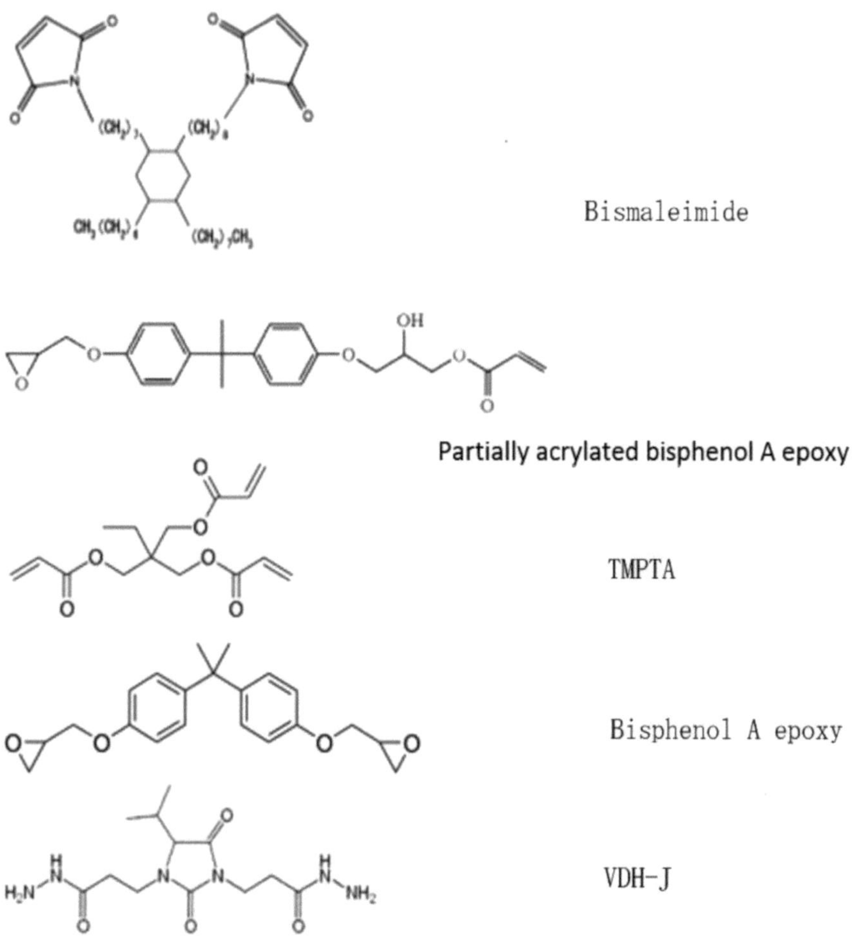

Figure 4. Chemical structure of reactive materials used.

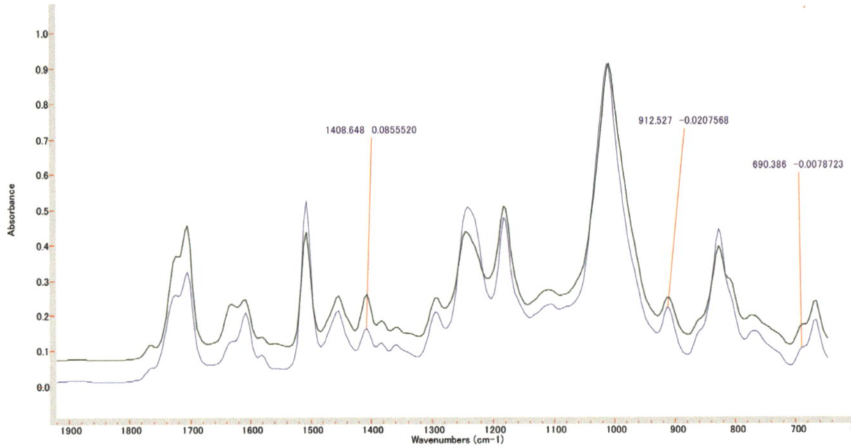

Figure 5. FT-IR spectrum of adhesive cured at UV only condition, in blue, compared to non-cure sample.

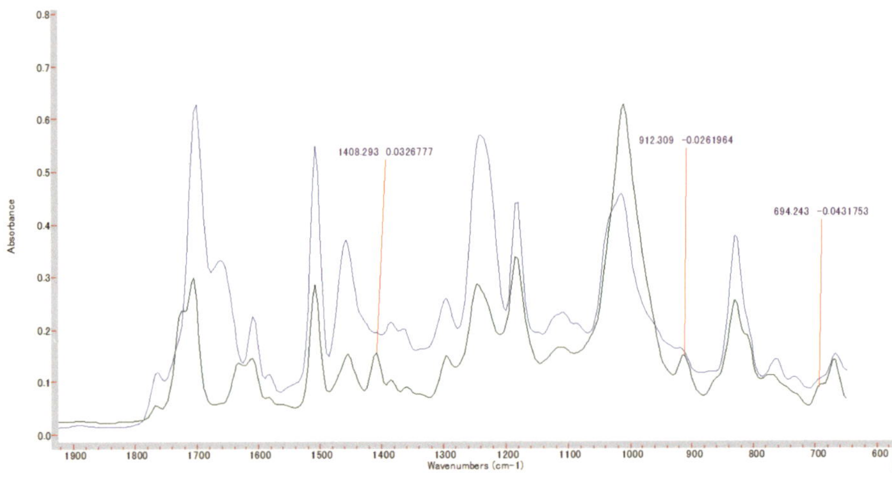

Figure 6. FT-IR spectrum of adhesive cured at UV + thermal cure condition, in blue, with compared to non-cure sample.

Cure condition	C=C conversion rate (%)		Epoxy conversion rate (%)
	Acrylate	Bismaleimide	
UV cure only, 100 mW/cm^2 × 30 s	62	95	0
UV + thermal cure, 100 mW/cm^2 × 30 s + 120°C × 60 min	100	96	85
Thermal cure only, 120°C × 60 min	67	95	69

Table 2. Conversion rate of C=C group and epoxy group measured by FT-IR.

the adhesive sample does not contain any thermal initiator component such as peroxide. Nevertheless, UV cure components of the adhesive sample showed also very good thermal curability. From epoxy resin part, conversion rate of epoxy group of adhesive sample cured at thermal cure only condition was lower than that cured at UV + thermal cure condition.

Based on this result, acrylic and maleimide double bonds reacted most probably with the epoxy curing agent, dihydrazine.

5. Summary

UV and thermal cure epoxy adhesives have been successfully used in high-end applications such as optical component bonding, sensor packaging and display panel assembly where high production efficiency and high adhesion performance are required. There are mainly two types of commercialized UV and thermal cure epoxy adhesives: UV cure cationic epoxy adhesives and hybrid UV acrylate and thermal cure epoxy adhesives. UV cure cationic epoxy adhesives are primarily composed of cycloaliphatic epoxy resin and cationic photoinitiator. UV cationic epoxy adhesives have no surface cure issue and possess low cure shrinkage and good adhesion performance but need post-thermal cure to achieve full adhesion performance. Hybrid UV acrylate and thermal cure epoxy adhesives are primarily composed of acrylate monomer, free radical photoinitiator, epoxy resin and curing agent. The hybrid epoxy adhesives combine fast UV curability of acrylate composition and high adhesion performance of thermal cure epoxy composition. A new type initiator free hybrid one-component UV and thermal cure adhesive has been also introduced. It is mainly composed of maleimide compound, acrylic monomer, partially acrylated epoxy resin, epoxy resin and latent curing agent. The new hybrid epoxy adhesive possesses good UV curability and satisfactory thermal curability and is suitable for use as high performance required applications.

Author details

Chunfu Chen[1]*, Bin Li[2], Chao Wang[2], Shuichi Iwasaki[1], Masao Kanari[1] and Daoqiang Lu[2]

*Address all correspondence to: chunfu.chen@jp.henkel.com

1 Henkel Technology Center—Asia Pacific, Henkel Japan Ltd., Yokohama, Japan

2 Henkel Adhesive Innovation Center, Henkel (China) Co., Ltd., Shanghai, PR China

References

[1] Petrie EM. Handbook of Adhesives and Sealants. MicGraw-Hill; 2006. 355p

[2] Sancaktar E, Bai L. Electrically conductive epoxy adhesives. Polymer. 2011;**3**:427-466. DOI: 10.3990/polym3010427

[3] Severijin C, Teixeira de Freitas S, Poulis JA. Susceptor-assisted induction curing behavior of a two component epoxy paste adhesive for aerospace applications. International Journal of Adhesion and Adhesives. 2017;**75**:155-164. DOI: 10.1016/j.ijadhadh.2017.03.005

[4] Vidil T, Tournilhac F, Musso S, Robisson A, Leibler L. Control of reactions and network structures of epoxy thermosets. Progress in Polymer Science. 2016;**62**:126-179. DOI: 10.1016/j.progpolymsci.2016.06.03

[5] Zotti A, Zuppolini S, Zarrelli M, Borriello A. Fracture toughening mechanisms in epoxy adhesives. In: Adhesives—Applications and Properties. London: InTech; 2016. pp. 237-269. DOI: 10.57772/65250

[6] Lewis AF. Epoxy resin adhesives. In: May CA, editor. Epoxy Resins—Chemistry and Technology. 2nd ed. New York: Marcel Dekker; 1988. p. 653

[7] Jin F-L, Li X, Park S-J. Synthesis and applications of epoxy resin: A review. Journal of Industry and Engineering Chemistry. 2015;**29**:1-11. DOI: 10.1016/j.jiec.2015.03.026

[8] Groulding TM. Epoxy resin adhesives. In: Pizzi A, Mittal KL, editors. Handbook of Adhesive Technology. 2nd ed. New York: Marcel Dekker; 2003. pp. 809-824

[9] Petrie EM. Epoxy Adhesive Formulations. New York: McGraw-Hill. p. 2006

[10] Thomas R, Sinturel C, Thomas S, El Akiaby EMS. Introduction. In: Thomas S, Sinturel C, Thomas R, editors. Micro- and Nanostructured Epoxy/Rubber Blends. Berlin: Wiley-VCH Verlag; 2014. p. 3

[11] Heise MS, Martin GC. Curing mechanism and thermal properties of epoxy-imidazole systems. Macromolecules. 1989;**22**:99-104

[12] Chen C, Iida K. Adhesives for flat-panel display manufacture. In: Adhesives, 2. Applications in Ullmann's Encyclopedia of Industry Chemistry. Berlin: Wiley-VCH Verlag; 2010. p. 519

[13] Herold J, Kluke M. UV light-curing adhesives for increased productivity. Radtech Report. 2012;**3**:27-31

[14] Sangermano M, Razza N, Crivello JV. Cationic UV-curing: Technology and applications. Macromolecular Materials and Engineering. 2014;**299**:775-793. DOI: 10.1002/mame.2013 00349

[15] Javadi A, Shokouhi H, Sobani M, Soucek MD. Cure-on-command technology: A review of the current state of the art. Progress in Organic Coatings. 2016;**100**:2-31. DOI: 10.1016/j.porgcoat.2016.02.014

[16] Lee C-S, Fan S, Seghier Z, Boey FYC, Abadie MJM. Photoreactivity of epoxy resins. Macromolecules. 2007;**3**:84-90

[17] Voytekunas VY, Ng FL, Abadie MJM. Kinetics study of the UV-initiated cationic polymerization of cycloaliphatic diepoxide resins. European Polymer Journal. 2008;**44**:3640-3649. DOI: 10.106/j.eurpolymj.2008.08.043

[18] Golaz B, Michaud V, Leterrie Y, Manson J-AE. UV intensity, temperature and dark-curing effects in cationic photo-polymerization of a cycloaliphatic epoxy resin. Polymer. 2012;**53**: 2038-2048. DOI: 10.1016/j.polymer.2012.03.025

[19] Jui-Hsun L, Youngblood JP. Adhesive bonding of carbon fiber reinforced composite using UV-curing epoxy resin. Composites Part B Engineering. 2015;**82**:221-225. DOI: 10.1016/j.compositesb.2015.08.022

[20] Atif M, Bongiovanni R, Yang J. Cationically UV-cured epoxy composites. Polymer Reviews. 2015;**55**:90-106. DOI: 10.1080/15583724.2014.963236

[21] Corcione C, Malucelli G, Frigione M, Maffezzoli A. UV-curable epoxy systems containing hyperbranched polymers: Kinetics investigation by photo-DSC and real-time FT-IR experiments. Polymer Testing. 2009;**28**:157-164. DOI: 10.1016/j.polymertesting.2008.11.002

[22] Kong S. Composition of cationic initiator and oxetane compound. US Patent 7902305

[23] Kong S, Grieshaber SE. Radiation or thermally curable barrier sealants. US Patent 8278401

[24] Hoshino T, Goto Y, Yada K. Resin composition. US Patent application 20150210905

[25] Gan Y, Chen C, Terada K. Cationically photocurable epoxy composition. US Patent 7456230

[26] Chiang TH, Hsieh T-E. A study of monomer's effect on adhesion strength of UV-curable resins. International Journal of Adhesion and Adhesives. 2006;**26**:520-531. DOI: 10.1016/j.ijadhadh.2005.07.004

[27] Chen C, Gan Y. Cationically curable epoxy composition. US Patent 7795744

[28] Velankar S, Pazos J, Cooper SL. High-performance UV-curable urethane acrylates via deblocking chemistry. Journal of Applied Polymer Science. 1996;**62**:1361-1376

[29] Fourassier J-P, Lalevee J. Photoinitiator for Polymer Synthesis. Berlin: Wiley-VCH Verlag; 2012. p. 41

[30] Ebnesajjad S. Adhesive Technology Handbook. 2nd ed. New York: William Andew; 2008. p. 124

[31] Fakley ME. Radiation-cured adhesives. In: Packham DE, editor. Handbook of Adhesion. John Wiley & Sons; 2005. p. 395

[32] Dekker C. UV-radiation curing of adhesives. In: Cognard P, editor. Adhesives and Sealants. Elsevier; 2002. p. 303

[33] Allen NS. Photoinitiators for UV and visible curing of coating: Mechanisms and properties. Journal of Photochemistry and Photobiology A. 1996;**100**:101-107

[34] Matsuda M. Sealants for one drop fill (ODF) process. In: Koide N, editor. The Liquid Crystal Display Story. New York: Springer; 2014. p. 199

[35] Park C, Lee S, Park J, Kim H. Preparation and characterization of dual curable adhesives containing epoxy and acrylate functionalities. Reactive & Functional Polymers. 2013;**73**: 641-646. DOI: 10.1016/j.reatfunctpolym.2013.01.012

[36] Xiao M, He Y, Nie J. Novel bisphenol A epoxide-acrylate hybrid oligomer and its photopolymerization. Designed Monomers and Polymers. 2008;**11**:383-394. DOI: 10.1163/156855508X3 32522

[37] Park Y, Lim D, Kim H, Park D, Sung I. UV- and thermal-curing behavior of dual-curable adhesives based on epoxy acrylate oligomers. International Journal of Adhesion and Adhesives. 2009;**29**:710-717. DOI: 10.1016/j.ijadhadh.2009.02.001

[38] Su Y, Cheng L, Cheng K, Don T. Synthesis and characterization of UV- and thermos-curable difunctional epoxy acrylates. Materials Chemistry and Physics. 2012;**132**:540, 100-549

[39] Sitmann E, Fuchs A, Worstatzky D. Photoinitiator: Their mechanism, use and applications. In: Florio JJ, Miller DJ, editors. Handbook of Coating Additives. Oxford: Taylor & Francis; 2004. pp. 61-126

[40] Chen CF, Iwasaki S, Kanari M, Li B, Wang C, Lu D. High performance UV and thermal cure hybrid epoxy adhesive. IOP Conference Series: Materials Science and Engineering. 2017;**213**:012032. DOI: 10.1088/1757-899X/213/1/012032

[41] Chen C. Sealing agent for liquid crystal dropping technology and method of manufacturing liquid crystal display. Japan Patent 5592081

[42] Chen C. Sealant composition. Japan Patent 5845341

[43] Bongiovanni R, Sangermano M, Malucelli G, Priola A. UV curing of photoinitiator-free systems containing bismaleimides and diacrylate resins. Progress in Organic Coatings. 2005;**53**:46-49. DOI: 10.1016/j.porgcoat.2004.11.009

[44] Vazquez GP, Joly-Duhamel C, Boutevin B. Photopolymerization without photoinitiator of bismaleimide-containing oligo(oxypropylene)s. Macromolecular Chemistry and Physics. 2009;**210**:269-278. DOI: 10.1002/macp.200800510

[45] Kuang W, Sabahi M, Nguyen C. Maleimide Reactive Oligomer for Wood Coating. RadTech Technical Proceeding. 2004. e/5

[46] Gonzalez MG, Cabanelas JC, Baselga J. Applications of FTIR on epoxy resins—Identification, monitoring the curing process, phase separation and water uptake. In: Theophile T, editor. Infrared Spectroscopy—Materials Science, Engineering and Technology. London: InTech; 2012. pp. 261-284

[47] Cholake ST, Mada MR, Raman RKS, Bai Y, Zhao X, Rizkalla S, et al. Quantitative analysis of curing mechanism of epoxy resin by mid- and near- FT-IR spectroscopy. Defence Science Journal. 2014;**64**(3):314-321. DOI: 10.14429/dsj.64.7326

[48] Ohtsuka K, Kimura H, Ikeshita S, Nakao H, Tsubota S. Novel bismaleimide/diallylbisphenol A resin modified with multifunctional thiol containing isocyanuric ring and long-chain aliphatic unit. High Performance Polymers. 2015;**28**(5):591. DOI: 10.1177/0954008315591191

Chapter 6

Transport of Electrolyte in Organic Coatings on Metal

Zoi Manoli, Darja Pecko, Guy Van Assche,
Johan Stiens, Ali Pourkazemi and Herman Terryn

Additional information is available at the end of the chapter

http://dx.doi.org/10.5772/intechopen.81422

Abstract

Organic coatings form an effective barrier between metals and their environment, providing them protection against corrosion. Corrosion on coated metals depends mainly on the diffusion of water through the coating, the loss of adhesion at the interface between the coating and the metal (delamination), the rate of the chemical and electrochemical reactions under the coating and the treatment of the metal surface before the coating application. Many aggressive ions are transported toward and inside the coating through water. In organic coatings, typically, the water absorbed by the coating affects the polymer matrix structure, and it causes swelling and stresses, which may result in cracks. Swelling and cracks enhance the transport of water into the solid polymer, and concurrently the diffusion of ions. Over time also, the chemical structure of the polymer may change, adversely affecting its barrier properties and overall performance. In this chapter, we focus on methods to quantify the transport of electrolyte in organic coatings. We mark out the main characteristics, advantages and limitations of each one of them.

Keywords: water uptake, organic coating, corrosion, gravimetric method, ATR-FTIR, NMR, THz-TDS, THz-FDS, EIS

1. Introduction

Organic coatings are used to protect metals against corrosion. In transportation, infrastructure, and oil and gas industry, just to name a few, corrosion has not only a serious economic impact (productivity and efficiency reduction), but it also harms the environment and reduces public safety. Therefore, coatings are extensively used to slow down the degradation of metals, thus extending the service lifetime of structures. An organic corrosion-protection coating applied on a metallic surface forms a barrier between the metal surface and its environment, preventing or delaying the transportation of corrosive species toward the metal, thus retarding the creation of rust at the metal surface. Over time, also the coating will degrade, reducing its barrier properties and corrosion protection efficiency. Scratches in the coating, pinholes, and surface contamination in between the coating and the metal are the main factors of coating degradation [1, 2]. Mechanical damages can be the pathways for enhanced water diffusion. The long-term performance of polymeric coatings highly depends on water adsorption and absorption. Typically, the polymer absorbs water, which affects the structure of the polymer matrix, causing swelling and internal stresses, thus producing nanosized channels and even cracks, enhancing the transport of water into the solid polymer and thus reducing its barrier properties. Additionally, the sorption of water may affect the polymer's mechanical properties by acting as a plasticizer and, in some cases, by causing hydrolysis. Progressively, the polymer degrades, losing its mechanical performance. The higher the temperature in a humid environment and the longer the time, the faster the polymer's degradation takes place. By calculating the amount of water that ingresses into the coating, and the diffusion coefficient, it is feasible to estimate the starting point in time where the corrosion may start occurring on the metal surface. After that point, the deterioration of the coating-metal interface begins. Given such parameters (amount and diffusion coefficient of water) it is feasible to optimize the quality of the coating (components, structure and thickness) in order to slow the process of corrosion down. The water uptake by coatings has been investigated by many researchers. However, most of the experimental techniques used, work under ideal and stable conditions, i.e. coatings without pinholes, scratches or any kind of defects, constant temperatures and humidity. The reason is that a combination of different physical and chemical phenomena can occur simultaneously; hence, different parameters affect the water uptake phenomena. The most common techniques to study water uptake in coatings are gravimetric method (GM), attenuated total reflection Fourier transform infrared spectroscopy (ATR-FTIR) and nuclear magnetic resonance (NMR). The amount of water sorpted can also be estimated using electrochemical techniques, such as electrochemical impedance spectroscopy (EIS). A novel, non-destructive evaluation for water penetration is provided by terahertz spectroscopy. Nevertheless, these techniques give reliable results given that they deal with ideal organic coatings (no defects, homogeneous surface) so as to consider only the diffusion phenomenon. In this chapter, we will first discuss the transport of water and electrolyte into organic coatings. Subsequently the experimental techniques for studying these processes will be looked at.

2. Organic coatings: barrier to electrolyte entrance

Organic coatings are thin solid-phase materials, which can be applied on structures to provide corrosion protection, and provide many other functions, such as, mechanical and chemical protection, electrical and thermal conductivity isolation, hydrophilicity or hydrophobicity, UV resistance, or esthetic properties, such as color, gloss or roughness [3, 4]. The formulation used to apply an organic coating typically consists of a binder and fillers, and in some case solvents or water (e.g., in latex paints). The compatibility of the components and their volume fraction, as well as the number of coating layers that are applied on a substrate, affects the properties and the functionality that the coating will finally reach. According to Szauer [5], an improper dispersion of molecules may leave free spaces available to be covered by corrosive species [6]. Despite development in coating technologies, troubles in corrosion protection of metals continue to exist in the long run. Substantial research has been done to understand scientifically the modes of failure of these materials in particular environmental conditions. Nevertheless, there has been limited work to quantitatively mark out the deterioration process and to predict the lifetime of the coated metal. One of the main reasons is the complexity of the coating/metal system and the combination of factors that simultaneously affect the performance of the coating. This encompasses the physical and chemical properties of the polymer, the substrate surface, the interface between coating and metal, the presence of defects, or local inhomogeneities. For example, low-molecular-weight or low-crosslinked materials can take up a large amount of water, and may show a low resistance to ion transport, hence they are susceptible to water attack. Corrosion takes place at the defects (pinholes, pores, scratches) and delamination or blisters appear adjacent to them.

When the coated metal is exposed to an aqueous solution, the water immediately starts spreading through the organic matrix. As for ion diffusion, most coatings work as a barrier due to their low dielectric constants and the limited free volume of their molecular structure. Nevertheless, once there is a defect in the organic coating or once there is enough water inside the polymer matrix, there is a transport link that connects the aqueous phase with the coating and the metal oxide film, through which water and hydrated ions migrate. This is the starting point for the deterioration of the coating first, and then of the oxide film and progressively of the metallic structure overall. The water molecules move toward the metal surface and they change the polymer properties. At the metal surface, they may replace the polymer and interact with the metal oxide film by opening a path for charged species [7, 8]. After the interaction with the metal oxide, electrochemical processes may start. With the creation of corrosion products underneath the coating, osmosis may be induced, creating mechanical forces that tend to separate the polymeric layer from the metal substrate [9] and cause local deformations [10, 11].

The diffusion of water and ions into the coating can be described in first approximation using Fick's law of diffusion, which describes these phenomena for a steady state (Eq. (1)):

$$J = -D\frac{(\partial c)}{(\partial x)} \tag{1}$$

where J is the diffusion flux, D is the diffusion coefficient, c is the concentration and x is the direction of the diffusion. To predict the water accumulation rate at non-steady state, Fick's second law of diffusion can be used (Eq. (2)):

$$\frac{\partial c}{\partial t} = D\left(\frac{\partial^2 c}{\partial x^2}\right) \tag{2}$$

where t is the diffusion time. This equation considers the coating as a homogeneous layer, where the diffusion takes place in one dimension and the diffusion coefficient is independent of distance, concentration and time. The temperature dependence of the Fickian diffusion of water is supposed to follow an Arrhenius model, as long as the temperature is below the glass

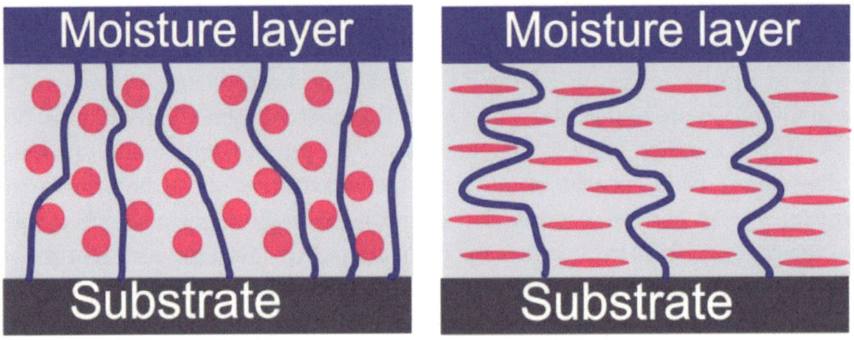

Figure 1. Spherical pigments allow easy penetration of water while lamellar pigments create more tortuous thus longer pathways for diffusion [23].

transition temperature (T_g) of the polymer. However, there are cases in which the transport of penetrants swells the polymer and changes the sorption kinetics, resulting in a non-Fickian diffusion process. This is common in more hydrophilic polymers, where the penetration of water plasticizes the polymer matrix. If this decreases the T_g (of the amorphous fraction) of the matrix sufficiently to have a change from a glassy to a rubbery state, diffusion may become 2–3 orders of magnitude faster. In addition, temperature variations during absorption, can permanently and significantly alter sorption behavior. The diffusion of water molecules in some organic coatings (e.g. epoxy) has been approached either volumetrically or interactionally [12]. The volumetric approach is based on the diffusion of water molecules through the free volume that present in between the macromolecular chains or the chain segments of the crosslinked network [13–16]. The free volume and the diffusion phenomenon depend on the crosslinking density and the physical state of the polymeric matrix. The nature of the penetrants (size, shape and polarity) influences the permeability and the rate of transport within the matrix. It has been reported that an increase in the size of the penetrant molecules causes a decrease in the diffusivity [17–19]. On the contrary, the interactional approach is based on the interactions between the polar groups and the water molecules, considering in general, the chemical nature of the components [20–22]. In general, the diffusion coefficient depends on the chemical functional groups and of the micro-organization of the polymer (void spaces, crosslinking and crystallinity). The presence of fillers and additives in the coating influences the chemical nature hence the diffusion coefficient too. **Figure 1** is an illustration of how the shape of the pigments in the coating may influence the pathway through which the water will manage to reach the substrate [23].

3. Experimental measurements of water sorption and transport

When a dry coating is exposed to an aqueous solution, water and ions slowly permeates the coating structure. There are many techniques to assess the transport properties of electrolyte in the coating film. In this section, we present the most common experimental methods to study the water uptake in organic coatings. There are two basic types of analysis: those which try to calculate the diffusion rates and those trying to demonstrate the concentration profiles. The most classic techniques are the gravimetric method (GM), attenuated total reflection Fourier transform infrared spectroscopy (ATR-FTIR), nuclear magnetic resonance (NMR) and electrochemical impedance spectroscopy (EIS). At the end of this section, we introduce a novel, non-destructive technique for evaluation of water penetration in coatings: terahertz spectroscopy (time-domain (TDS) and frequency-domain (FDS)).

3.1. Gravimetric method (GM)

The gravimetric method is probably the oldest method to determine the sorption of water by a coating. It is still widely used because it is a simple and accurate way by measuring the total weight gain after exposing the coating to liquid or vapor for a period of time. Different (weighing) equipment is used depending on the sensitivity required. The sample, usually a coated metal substrate or a free-standing coating, is immersed in the electrolyte and removed for weighing at different points of time. It is essential to wipe the excess liquid from the exposed surface before placing the sample on a microbalance. The weight gain (Eq. (3)) at time t, assuming a constant penetration of liquid in an initially completely dry film of thickness L, is described by Crank and Park [24]:

$$\frac{M_t}{M_\infty} = 1 - \frac{8}{\pi^2}\sum_{n=0}^{\infty}\frac{1}{(2n+1)^2}\, exp\left[\frac{-D(2n+1)^2\pi^2 t}{L^2}\right] \tag{3}$$

where M_t is the absorbed mass at time t, M_∞ is the equilibrium mass and D is the diffusion coefficient that is assumed to be constant. This formula considers that the uptake of solute into polymers is following Fick's law. Kloppers et al. [25] investigated the water uptake in PET films of different thickness. **Figure 2** shows that the water diffusion is Fickian and that the relative mass change M_t/M_∞ versus the dimensionless time $\tau = tD/L^2$ plot, can be described by the Eq. (3).

3.2. Attenuated total reflection Fourier transform infrared spectroscopy (ATR-FTIR)

The diffusion of water in an organic coating can also be followed using attenuated total reflection Fourier transform infrared spectroscopy (ATR-FTIR) [26, 27]. Infrared spectroscopy provides quantitative information on the composition of complex molecules or mixtures of molecules, by studying the absorption of IR light resulting from vibrations of functional groups. As typically the thickness of a polymer film studied in transmission needs to be less than 10 μm to avoid complete absorption in sections of the spectral region, often ATR-FTIR is used for thin coatings. In ATR-FTIR, the IR beam passes through the ATR crystal (often germanium, silicon, ZnSe or diamond) and reflects at the interface between the crystal and the sample. This reflection creates an evanescent wave that penetrates into the coating to a depth of a few micrometers. The penetration depth depends mainly on the refractive indices of the ATR crystal and the material of the sample, the wavelength of the light beam, and the angle of incidence. The refractive index of the ATR crystal must be greater than the refractive index of the sample. The beam exits the crystal and it is collected by a detector, giving an ATR infrared spectrum. This technique allows for following the water sorption by an organic

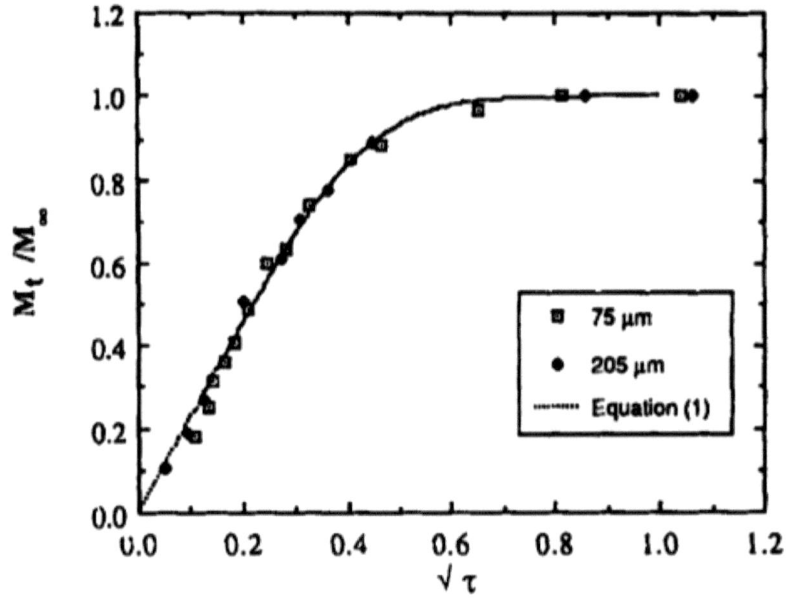

Figure 2. Water uptake against the square root of the dimensionless time, τ, for the PET films of 75 and 205 μm thickness and the theoretical result taken from Eq. (3) [25].

coating when it is exposed to an aqueous solution. Nguyen et al. [28] demonstrated a set-up that detects the water at the coating/substrate interface regardless the thickness of the coating and Fieldson and Barbari [26] and other researchers studied the water sorption kinetics in polymeric films [29]. Typically, the sample is a free-standing organic coating that is placed between the ATR crystal and the diffusing medium. With this method it is possible to measure the amount of absorbed water as a function of time, in experiments taking place in situ, by recording time-resolved IR spectra of the hydroxyl (νO–H stretch) absorption band, typically in the range of 3000–4000 cm^{-1}. It is essential that the coating is in direct contact with the crystal. Recently, several polymeric materials have been analyzed for water sorption and desorption in this way [30–34]. Wapner et al. [35, 36] used a similar experimental set-up as in **Figure 3**, to measure the uptake of water and ion transport at polymer/metal interface. They used deuterated water instead of normal water because the IR-Bands of the former do not overlap with the bands from the epoxy resin.

The principle behind the quantitative analysis of the spectra is the direct relationship between the absorption of electromagnetic waves and the quantity of the absorbing material. For transmission FTIR, the Beer–Lambert law represents this relationship (Eq. (4)):

$$dI = -\varepsilon \ln 10\, cI\, dz \tag{4}$$

where I is the light intensity at position z, ε is the molar absorption, c is the concentration of absorbing substance. By integrating Eq. (4), the concentration profile over the thickness of the film is accounted for (Eq. (5)):

$$A = -\log\frac{I}{I_0} = \int_0^L \varepsilon c\, dz \tag{5}$$

where A is the measured absorbance, I_0 is the intensity of the incident light, I is the intensity of the transmitted light, L is the thickness of the coating, over which there is the absorbing group.

In ATR-FTIR, the evanescent wave field decays exponentially in the less medium according to Eq. (6), and the penetration depth of the evanescent IR beam into the sample is the reciprocal of Eq. (7):

$$E = E_0 e^{-\gamma z} \tag{6}$$

$$\gamma = 2n_2\pi\frac{\sqrt{sin^2\varphi - \left(\frac{n_1}{n_2}\right)^2}}{\lambda} = \frac{1}{d_p} \tag{7}$$

where E_0 is the electrical field strength at the interface, I equals E^2, z is the distance from the surface, n_1 is the refractive index of ATR crystal and n_2 is the refractive index of the coating, φ is the angle of incidence of the infrared beam and d_p is the depth of penetration of the IR beam into the sample. Assuming Fickian diffusion, the diffusion coefficient can be calculated by the equation that describes the intensity at a given point as a function of the electrical field (Eq. (8)):

$$\frac{A_t}{A_\infty} = 1 - \frac{8\gamma}{\pi\left[1 - \exp\left(-2L\gamma\right)\right]}\left[\frac{\exp\left(-\frac{D\pi^2 t}{4L^2}\right)\left(\frac{\pi}{2L}\, exp\left(-2\gamma L\right) + \left(2\gamma\right)\right)}{\left(4\gamma^2 + \frac{\pi^2}{4L^2}\right)}\right] \tag{8}$$

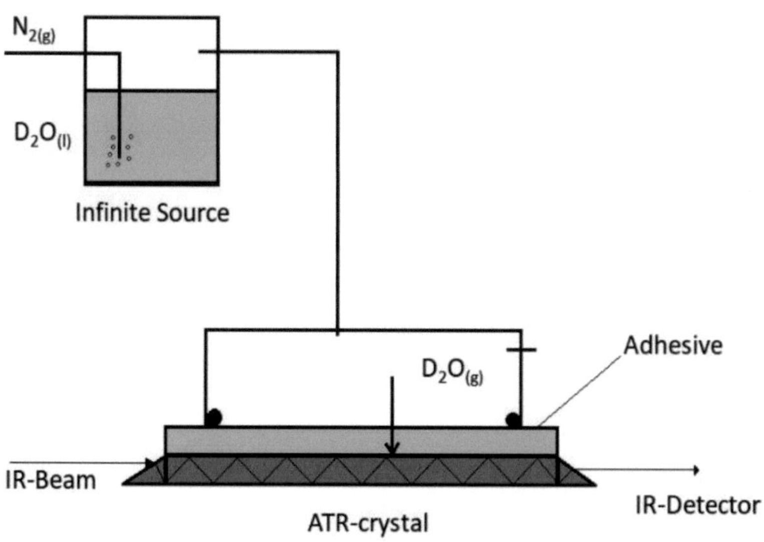

Figure 3. Experimental set-up for the measurement of water diffusion using ATR-FTIR. The deuterated water was used instead of water as the IR-Bands of the deuterated water do not overlap with bands from the epoxy resin [35].

where A_t is the absorbance at time t, A_∞ is the absorbance at equilibrium, and D is the diffusion coefficient [25]. In the IR spectrum, each stretching and bending vibration occurs with a characteristic frequency as the atoms and charges involved are different for different bonds. Typical difference ATR-FTIR spectra for the water absorption in an alcohol/amine adhesive film exposed to an atmosphere saturated with deuterated water (D_2O) for different exposure times has been demonstrated by Wapner et al. [35] in **Figure 4**. They also calculated the diffusion coefficient based on the increase of the area of the νOD-band and by fitting that curve with Eq. (8). The negative νO–H bands are because deuterated water molecules (D-protons) replaced the H-protons of the polymer and lowered the intensity of the νO–H absorption bands (the higher mass of the D atom shifts the wave number to lower values). According to Nguyen et al. [37], the idea behind the increasingly negative CH_x bands is that the water replaces the polymer at the interface and builds up an aqueous layer, leading to a gradual loss of adhesion. Given that penetration depth of the IR beam is nearly constant during an experiment, the lowering in the intensity of the polymer bands occurs as the polymer is pushed away by the developing aqueous layer.

The quantification of water at the coating/substrate interface is derived from the penetration depth concept of internal reflection spectroscopy developed for thin and thick films. Supposing that l is the thickness of the water layer at the coating/substrate interface; d_{pw} and d_{pc} are the penetration depths of the evanescent wave in water and coating, respectively; n_2 is the refractive index and α_2 is the absorption coefficient of water at the coating/substrate interface, then the thickness of the water is given by Eqs. (9) and (10):

$$l = \frac{d_{pw}}{2}\left[-\ln\frac{1-\frac{A}{A_\infty}}{1-c_w\frac{d_{pc}}{d_{pw}}}\right] \tag{9}$$

$$A_\infty = \frac{n_2\alpha_2E_0^2d_{pw}}{2n_1\cos\theta} \tag{10}$$

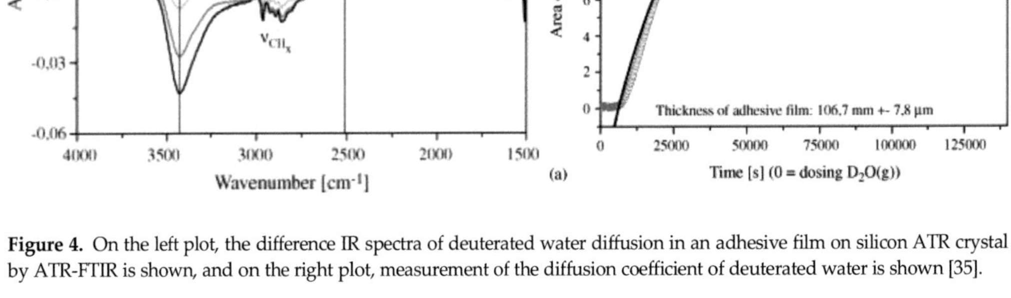

Figure 4. On the left plot, the difference IR spectra of deuterated water diffusion in an adhesive film on silicon ATR crystal by ATR-FTIR is shown, and on the right plot, measurement of the diffusion coefficient of deuterated water is shown [35].

where c_w is the fraction of water sorbed in the coating within the probing depth and A_∞ is the IR absorbance when the water layer at the coating/substrate interface is very thick. The amount of water Q is given by Eq. (11) where α is the area in contact with water and ρ is the density of water at the interface:

$$Q = l\alpha\rho \tag{11}$$

3.3. Nuclear magnetic resonance (NMR)

Nuclear magnetic resonance (NMR) is a spectroscopic technique based on the absorption of electromagnetic radiation in the radio frequency range (4–900 MHz). It is widely used to obtain structural information or even identify molecules in solution. Compared to IR spectroscopy, NMR can also provide information on molecular dynamics.

NMR takes advantage of the spin states of protons or other atoms with a non-zero spin. In a (static) magnetic field, the nuclei resonate at a specific resonance frequency between states with a magnetic moment parallel and opposite to the external field. Thus, the energy levels of the nucleus are split by the magnetic field, and nuclei can be excited from the lower to the higher level by a radio frequency (RF) pulse. More specifically, Eq. (12) shows the linear dependence between the resonance frequency and the magnitude of the applied magnetic field:

$$f = \frac{\gamma}{2\pi}|\vec{B}| \tag{12}$$

where f is the resonance frequency in MHz, γ is a constant and the ratio $\gamma/2\pi$ is the gyromagnetic ratio and it is different for different kinds of nuclei. $\left|\vec{B}\right|$ is the magnitude of the applied magnetic field in tesla (T) [38, 39]. For example, the hydrogen nucleus has a constant $\gamma = 42.58$ MHz/T. By exciting the nuclei with a radio frequency pulse at the resonance frequency, the nuclei in return start emitting signals recorded by a receiver coil. The most common way of extracting chemical information from NMR signals is the chemical shift: the applied magnetic field induces currents in the electron cloud surrounding the nucleus, creating an induced magnetic field of opposite sign, reducing the field sensed by the nucleus. This induced field has a magnitude between 10^{-4} and 10^{-6} times the size of the applied field.

NMR relaxometry can provide information about the diffusivity of liquids [40] and the mobility of the polymer molecules [41]. After giving an RF-pulse (e.g., a 90° or 180° pulse), the magnetization will relax to its equilibrium state through spin lattice and spin–spin relaxation, characterized by relaxation times T_1 and T_2, respectively, which are very sensitive to the molecular motion. For T_2 relaxation, a CPMG sequence is used, leading to a signal S described by the multiexponential decay given in Eq. (13):

$$S(t) = \sum_i S_i e^{-t/T_{2i}} \tag{13}$$

where the S_i is the component intensity, t is the time from the 90° pulse, and T_{2i} is the component relaxation time, which will increase with the mobility of the component. Thus, the components of the signal decay are correlated with coating components and their mobility. For polymers, what matters is the segmental mobility of the polymeric chains. Hence, plasticizing effects of sorbed water can be detected since this phenomenon leads to a mobility increase of the polymer chains [42]. For water, the T_2 relaxation time will depend on the state (bound or free) and on the pore size distribution [43, 44].

Glover, Aptaker et al. [45, 46] proposed the GARField approach, a powerful magnetic resonance imaging (MRI) method that can be used for monitoring the ingress of water in coatings and the corresponding changes in molecular mobility [47]. Baukh et al. [38] used this approach to visualize the water distribution in multilayer polymeric films: using a high static gradient of the magnetic field generated by special shaped magnetic poles, depth profiles over 500 μm could be analyzed with a 2.5–6 μm resolution. The experimental set-up is shown in **Figure 5**.

For a multilayered film combining a top coat and a base coat on glass, the intensities of NMR profiles taken from both wet and dry coatings were compared. The signal intensity of the wet coating was higher than the intensity of the dry one (**Figure 6**). The polymer was also exposed to D_2O, which is expected to behave very similar to water molecules. Progressively, D_2O molecules affect the polymer matrix mobility, and as deuterium cannot be probed, the only detected NMR signal is from the polymer. This way it is possible to follow the mobility of the polymeric chains, reflected by the enhancement of the NMR signal.

The signal increase $\Delta S = [S_{D2O/H2O}\text{-}S_{D2O}]$, where S are the integrated signals of the base coat layer in the case of exposure to a D_2O/H_2O mixture and to pure water, respectively, is proportional to the mass of water in the coating Δm (mg) and can be written as $\Delta S = k\Delta m$, where k (mg^{-1}), is a proportionality coefficient given by (Eq. (14)):

$$k = \frac{1}{\rho_w A}\left[1 - exp\left(-\frac{t_r}{T_1}\right) exp\left(-\frac{t_e}{T_2}\right)\right] \tag{14}$$

where ρ_w is the density of liquid water, A is the area of the base coat, T_1 and T_2 are the relaxation times of water, and where t_r is the repetition time and t_e is the echo time of the pulse sequence used. Using this approach, the increase in water content in the base coat due to

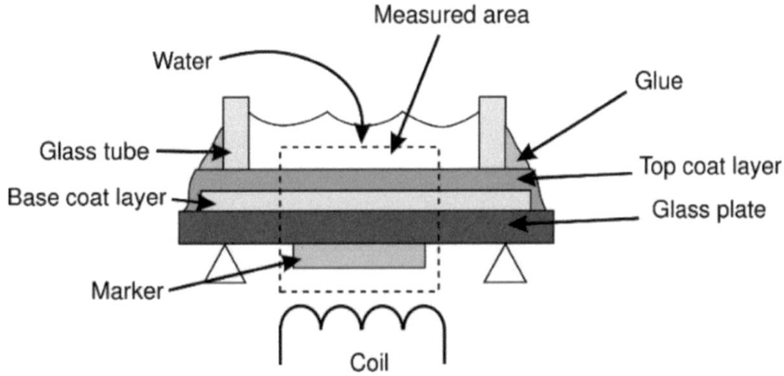

Figure 5. Schematic representation of the experimental set-up for the measurement of the water uptake using the GARField approach. The samples were put on top of the RF coil. To wet a sample, a glass tube was glued on top of the coating [38].

http://dx.doi.org/10.5772/intechopen.81422

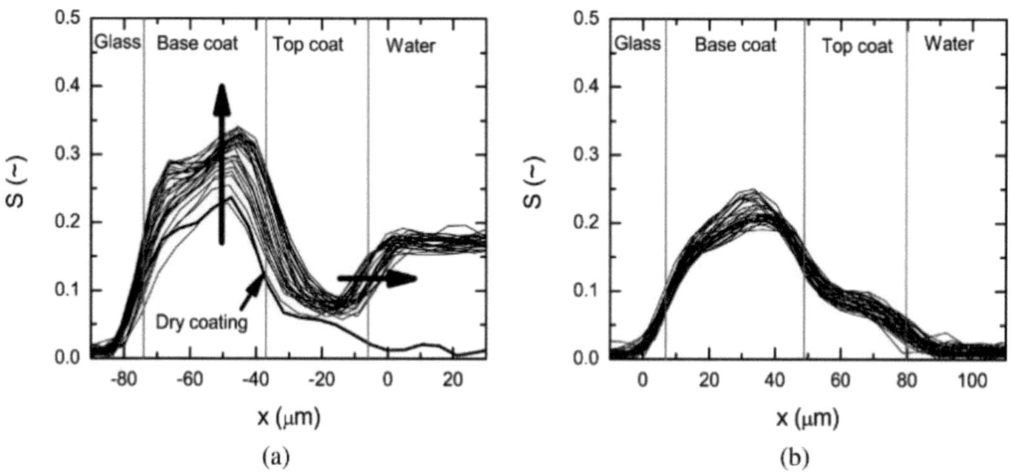

Figure 6. NMR signals profiles in the sample of base coat during (a) water and (b) heavy water uptake. The vertical arrow indicates the signal increase in the base coat and the horizontal arrow indicates swelling. The lower bold line in (a) represents the signal of the dry sample [38].

diffusion of water through the top coat can be followed, permitting an (indirect) estimation of the diffusion coefficient in the top coat.

If both diffusion and molecular mobility contribute to the relaxation process, the measured T_2 time will be given by (Eq. (15)):

$$\frac{1}{T_2} = \frac{1}{T_{2D}} + \frac{1}{T_{2S}} \tag{15}$$

where T_{2D} represents relaxation due to diffusion and T_{2S} relaxation corresponding to the mobility of the measured species. The relaxation due to diffusion in the field gradient is given by (Eq. (16))

$$T_{2D}{}^{-1} = \alpha\gamma^2 G^2 D t_e{}^2 \tag{16}$$

where D is the self-diffusion coefficient of the measured species, α is a constant defined by the evolution of coherent pathways for a given pulse sequence and is determined using the measurement of the diffusivity of pure water in a dilute aqueous reference solution, and G is the gradient of the magnetic field. Baukh et al. [48] estimated the self-diffusion coefficients of water in a fully and partially saturated base coat using the slope of the linear part of the $T_2{}^{-1}$ versus $t_e{}^2$ curve (Eq. (16)). For a more detailed explanation of the principles and equations, the reader is referred to the papers of Huinink and coworkers [38, 47, 48].

3.4. Terahertz spectroscopy: time-domain (TDS) and frequency-domain spectroscopy (FDS)

The frequency region of electromagnetic waves ranging from ~30 GHz to ~10 THz covers typically three bands, called as the mmw band, the sub-THz band and the THz-band which are situated between the microwave band and the infrared. In order to simplify the terminology related to this spectral range, we call this spectral part range the THz domain, which is the least exploited of the electromagnetic spectrum. Nevertheless, recently spectroscopies methods operating in this region have been used for the characterization of various materials. THz radiation does not readily penetrate through metals or pure polar liquids such as water. However, this limitation of THz propagation through water turns into an advantage in detecting moisture or humidity in materials as water has a large relative permittivity and is highly absorptive in the THz range and at the same time the considered organic materials which absorb moisture are reasonably transparent for the lower frequencies of the considered THz band. Hence, the contrast in the image between the wet and the dry regions is high. Normally, the water diffusion in organic coatings takes some hours or days. In general, THz imaging systems can operate in the frequency or the time domain, the latter one featuring broadband characteristics, allowing low spectral resolution spectroscopy and typically more expensive, the former one allowing much higher spectral resolution. THz imaging systems can scan the full coating area in a few to tens of seconds, therefore allowing to follow the water ingress in the coating from the very early stages of the diffusion. Another property of THz imaging is the fact that it can produce 2D images of the water ingress and detect structures comparable in size to the electromagnetic wavelength of the THz radiation (from 10 mm down to 0.3 mm). For example, it can spot water diffusion into comparably-sized voids in the subjected material as it will create a large contrast between the dry and the wet THz absorbance images [49]. The first demonstration of THz imaging for mapping the moisture has been done in drying leaves [50, 51]. Overall, this method enables the creation of images of liquid diffusion in materials and depicts possible diffusion pathways, the extraction of average diffusion coefficients, and the non-destructive assessment of hydration levels inside objects. The high permittivity of water is mainly the feature that influences the contrast mechanism for moisture detection by THz imaging. The ingress of water in the material is followed by an increase of the overall absorbance at all THz frequencies.

Jordens et al. [52] studied the water absorption in polyamide and wood plastic composite using THz spectroscopy. By comparison with gravimetry results, they concluded that the refractive index and the absorption coefficient obtained (at 600 GHz) can be employed to derive the volumetric water content of a polymer from the measured THz signal after calibration. Although the contrast in permittivity between the coating material and the water is high, for very thin layers (sub-100 μm) with very low water content (only a few %), sensitivity enhancement techniques (e.g. resonant like structures) are needed [53, 54]. These techniques can only be implemented in the frequency domain. Recently, Pandey et al. [55] showed that mm-waves (around 60 GHz) can be used to follow the drying or freezing of food slices with high sensitivity. However, all the mentioned methods are not completely blind and need priori information (e.g. the thickness of the coating) to extract the absolute water content. Using dedicated setups exploiting the transient radar method, also the blind analysis of multilayer structures with deep submillimeter depth resolution, significantly below the wavelength, is feasible [56]. This method allows to determine the dynamic thickness of the coating (e.g. due to swelling) and the frontline of the water ingress.

Obradovic et al. [57] studied the diffusion coefficient of acetone in polycarbonate and polyvinylchloride polymers by using a THz reflective geometry (**Figure 7**). **Figure 8**, on the left shows THz time domain measurements presented as a stack plot as a function of

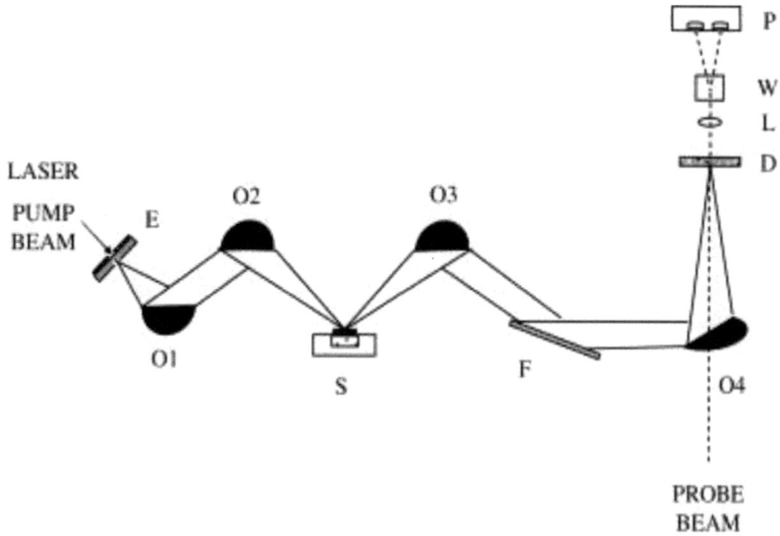

Figure 7. An illustration of the reflectance spectrometer. *E* is the emitter, *O* is the off-axis parabolic mirrors, *F* is the flat mirror, *S* is the sample holder with polymer disc, *D* is the detector, *L* is the lens, *W* is the Wollaston prism, and *P* is the photodiodes [57].

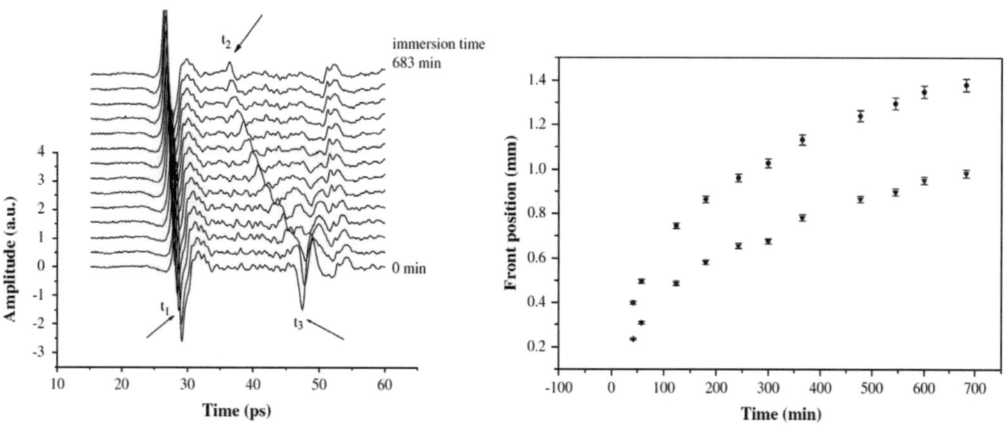

Figure 8. THz-time domain measurements of acetone ingress into PVC at room temperature. On the left the plot, the waveforms at time t_1 correspond to the top surface, at t_2, they come from the penetrant front interface and at time t_3, from the rear surface of the polymer. On the right plot, the acetone depth penetration into the polymer as a function to time is shown [57].

exposure time of the PVC to the acetone, and on the right, it shows the evolution of the depth of penetration of acetone versus time.

Although diffusion coefficients or quantitative diffusion models have not been determined yet using THz-TDS, the high sensitivity for water and the ability to measure with submillimeter depth resolution are promising for following concentration profiles during the ingress of water in coatings. Kusano et al. [58] studied the penetration of acid solutions into epoxy resins by THz-TDS. They found that the refractive index of epoxy specimens is changing as this is due to the solution uptake.

3.5. Electrochemical impedance spectroscopy (EIS)

Electrochemical impedance spectroscopy is an in situ, non-destructive technique that can be used to characterize a wide variety of electrochemical systems, including the degradation of the corrosion protection of coated metals over time. The EIS experiments are usually done in an electrochemical cell with three electrodes immersed in a liquid electrolyte: the working electrode, which is the coated metal, the counter electrode (made of an inert conductor, such as platinum) that provides a circuit along with the working electrode, and the reference electrode that is used to determine the potential of the working electrode (e.g. made of Ag/AgCl). The impedance is measured between the reference electrode and the working electrode. The electrochemical cell is connected to a potentiostat instrument that controls the voltage difference between the reference and the working electrode. Deflorian et al. [59] studied the ion diffusion through organic coatings with EIS, working in the electrochemical cell of **Figure 9**.

In general, EIS involves measuring the electrical response of a system subdued to a low amplitude, sinusoidal potential perturbation (Eq. (17)), typically in the range of ± 10 mV, as a function of frequency, usually between 10^{-2} and 10^{5} Hz. The response of the system is an alternating current signal with the same frequencies (the linear part of the response, Eq. (18)) and higher harmonics (the non-linear part):

$$E(t) = E_0 \cos(\omega t) \tag{17}$$

$$I(t) = I_0 \cos(\omega t - \varphi) \tag{18}$$

where E_0 is the amplitude of the potential difference, I_0 is the amplitude of the current, $\omega = 2\pi f$ is the angular frequency and t is the time, φ is the phase shift, time shift or phase angle, and it exists because the response depends on the physicochemical processes taking place. In

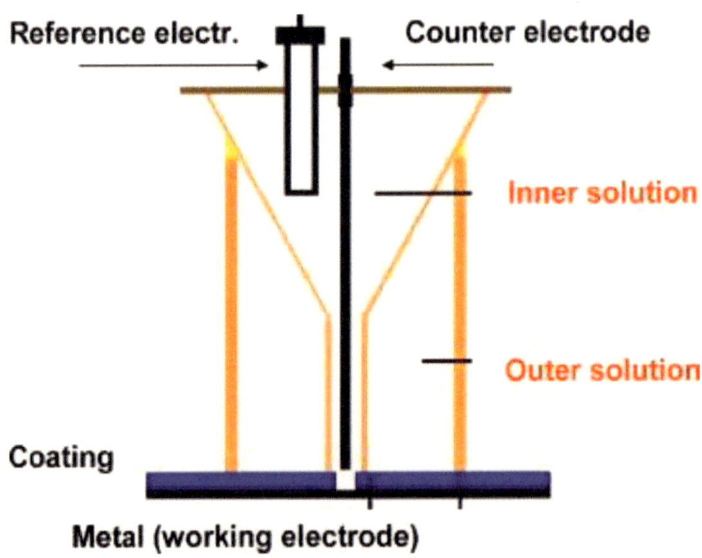

Figure 9. Electrochemical cell with three electrodes (working, counter and reference) for analyzing the diffusion of water into a coating [59].

corrosion studies, the current response is not linear due to polarization and/or passivation effects. However, for small perturbation signals, the response may behave as pseudo-linear.

Electrochemical processes that may take place in the coating-oxide-metal system can be described by equivalent electrical circuits consisting of a series of resistors, capacitors, and inductors, sometimes extended with special elements, like a Warburg element (when the system is dominated by mass transport phenomena). More specifically, the recorded impedance spectra are analyzed by fitting electrochemical equivalent circuits to the data. The change of certain parameters in the circuits can then be correlated with specific changes in the system over time, such as, an increasing permittivity of the coating that results from an increasing water sorption. The most simple electrical circuit would be a circuit with an electrical resistance only, defined by Ohm's law and mathematically expressed in Eq. (19):

$$R = \frac{E}{I} \tag{19}$$

where R is the resistance, E is the applied alternating potential difference, I is the alternating current response. For complex systems, the equivalent circuits get more complicated, and one needs to move from the real to the imaginary representation, introducing the concept of impedance (Z). By turning Eqs. (17) and (18) into complex functions, the potential (Eq. (20)) and the current (Eq. (21)) are described as follows:

$$\widetilde{\boldsymbol{E}}(\boldsymbol{t}) = \boldsymbol{E_0}\, \boldsymbol{exp}\left(\boldsymbol{j\omega t}\right) \tag{20}$$

$$\widetilde{\boldsymbol{I}}(\boldsymbol{t}) = \boldsymbol{I_0}\, \boldsymbol{exp}\left(\boldsymbol{j\omega t} - \boldsymbol{\varphi}\right) \tag{21}$$

where j is the complex number ($j^2 = -1$). The measured potential and current are the real parts of the imaginary functions. The impedance is a measure of an electrical circuit's opposition to an alternating current when an alternating potential is applied. In analogy with Ohm's law, the impedance Z is expressed as (Eq. (22)):

$$Z = \frac{\widetilde{E}(t)}{\widetilde{I}(t)} = Z_0\, exp\,(j\varphi) = Z_0(\cos\varphi + j\sin\varphi) \tag{22}$$

where the impedance Z is a complex function with magnitude or modulus Z_0 expressed in ohms. The impedance consists of a frequency dependent real and imaginary part, which can be plotted in a Nyquist plot (**Figure 10**). Alternatively, the magnitude Z_0 and the phase angle φ of the impedance can be plotted as a function of frequency, giving the Bode plot (**Figure 12**). The semicircle of **Figure 10** reflects the impedance of a Randle's circuit and is characteristic for a single time constant. For more complex circuits, the Nyquist plot may contain several semicircles and often only a portion of a semicircle can be seen.

One of the important electrical elements for modeling the electrolyte-coating-oxide-metal system is the capacitor. The classical capacitor is composed of a non-conductive medium (dielectric) between two conductive plates. The capacitance depends on the dielectric properties of the medium, on the thickness of the medium, and on the size of the plates. For coating evaluation, taking a coated metal immersed in an electrolyte, the water represents one plate

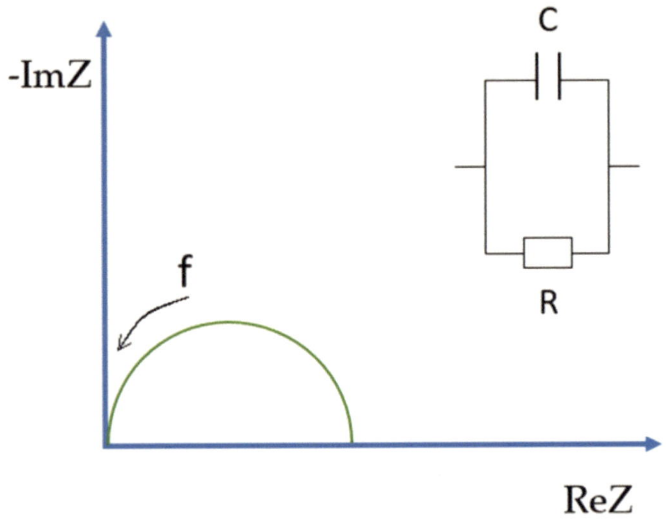

Figure 10. Nyquist plot for Randle's circuit made of a resistor in parallel with a capacitor.

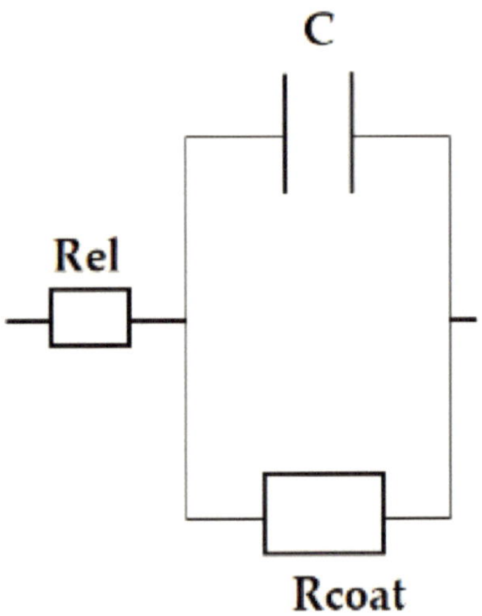

Figure 11. Equivalent circuit of coated metal system immersed in an electrolyte solution.

and the metal represents the other one, while the coating between the electrolyte and the metal can be considered as the non-conductive medium (the dielectric). Thus, EIS can be used to measure in situ the capacitance of the coating on metal system during exposure. The simplest equivalent electrical model that represents a coated metal immersed in an electrolyte is given by **Figure 11**, where R_{el} is the resistance of the electrolyte, C is the capacitance of the coating, and R_{coat} is the resistance of the coating.

The relation between the capacitance and the permittivity of the polymer is given by Eq. (23):

$$C = \frac{\varepsilon_r \varepsilon_0 A}{d} \tag{23}$$

where C is the capacitance derived from the impedance measurements (expressed in F), ε_r is the relative permittivity of the coating (dielectric constant), ε_o is the permittivity in vacuum (in F), A is the surface of the coated metal, and d is the thickness of the coating. Typically, the dielectric constant of organic coatings is between 2 and 7, whereas the dielectric constant of water is 80.1 at 20°C. The diffusion of water into a coating increases the polarisability, and hence the permittivity of the polymer. Therefore, the absorption of water by the coating causes an increase in the capacitance, making EIS suitable to follow the absorption of water in a coating [60]. The magnitude of the impedance measured, represented by the electrochemical circuit of **Figure 11**, tends to decrease and the phase angle starts to increase as the coating absorbs water. The coating is considered bad when the coating impedance at low frequencies drops below 10^7 Ωcm^2 (this is evaluated per cm^2 of coating exposed to the electrolyte). Fredj et al. [61] studied the water ingress in epoxy organic coatings by EIS. **Figure 12** shows the Bode plots of an epoxy coating immersed in an electrolyte for different times of immersion, showing the impedance magnitude and (minus) the phase angle as a function of frequency. The phase angle starts from -90°, reflecting the (predominantly) capacitive behavior of the initial coating. For long immersion times, the phase angle at low frequencies evolves to zero, indicating a (predominantly) resistive behavior at these frequencies.

According to Deflorian et al. [62] the ideal coating capacitance increases in three phases, as indicated in **Figure 13**. Phase I represents a situation where the electrolyte diffuses homogeneously into the coating and usually it is described mathematically by Fick's first law (Eq. (1)). In phase II, the plateau represents the water saturation in the polymer matrix, where the capacitance remains constant over time. Finally, phase III shows a further increase of the water inside the coating, most probably in a non-homogeneous way, due to (local) chemical changes in the coating.

The capacitance values derived using equivalent electrical models can be exploited to quantify the water volume fraction through the Brasher/Kingsbury [63, 64] equation (Eq. (24)):

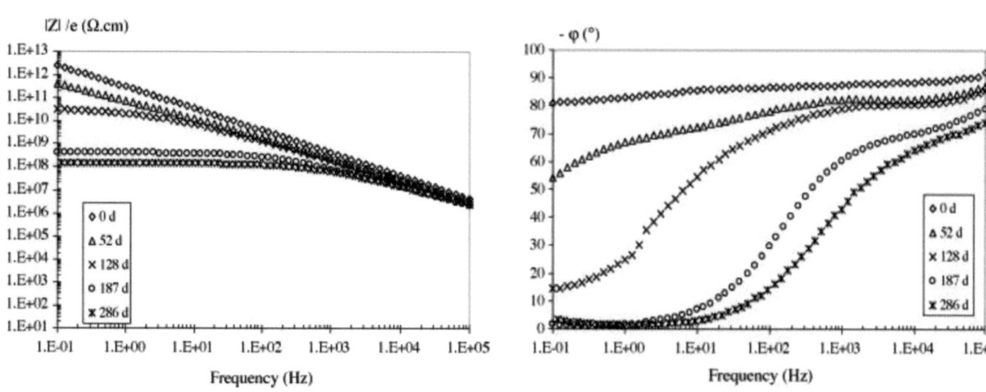

Figure 12. Bode plots for an epoxy coating immersed in sodium chloride solution for increasing times [61]. Note that the impedance shown here in ohm.cm, is the impedance measured per cm^2 coating surface and per cm coating thickness, which is suitable for evaluating the coating material.

$$V_t = 100\frac{\log \frac{C_m}{C_{m,o}}}{\log \varepsilon_w} \tag{24}$$

where V_t is the water volume % of water absorbed, C_m is the measured capacitance at time t, $C_{m,o}$ is the measured capacitance at time zero, and ε_w is the dielectric constant of water at the experiment temperature T. The use of Eq. (24) considers some of the following assumptions: the change in capacitance of the coating is only due to an increase in absorbed water, the water is well distributed throughout the coating, no swelling occurs due to the gradual water uptake, no interactions occur between water and polymer molecules. If these assumptions are not fulfilled, then there is a possibility of overestimating or underestimating the water content. By estimating the water content from the capacitance, it is also feasible to calculate the kinetics of water uptake (diffusion coefficient) [65, 66]. Starting from Fick's second law and the use of the value of the capacitance at different immersion times, the following formula is used (Eq. (25)) [67]:

$$\frac{C_m - C_o}{C_\infty - C_{m,o}} = \frac{4\sqrt{D}}{d\sqrt{\pi}}\sqrt{t} \tag{25}$$

where D is the diffusion coefficient, and C_∞ is the capacitance at saturation. In this approach, it is assumed that the water ingress follows a Fickian relation, although, as mentioned above, this is not always the case.

Working on coating films with EIS and the gravimetric method, and using a more complex equivalent circuit that includes constant phase elements (CPE), Nguyen et al. [68] suggested that in some cases the first step of the diffusion process can be described by Fick's law, but during the second step (water saturation state) the diffusion process is described by non-Fickian behavior due to swelling (**Figure 14**).

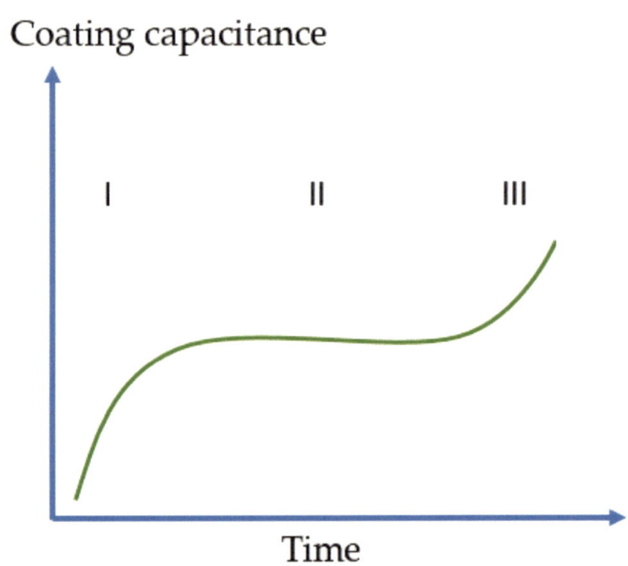

Figure 13. Coating capacitance evolution as a function of time for coatings exposed to an aqueous solution (*redrawn after* [62]).

http://dx.doi.org/10.5772/intechopen.81422

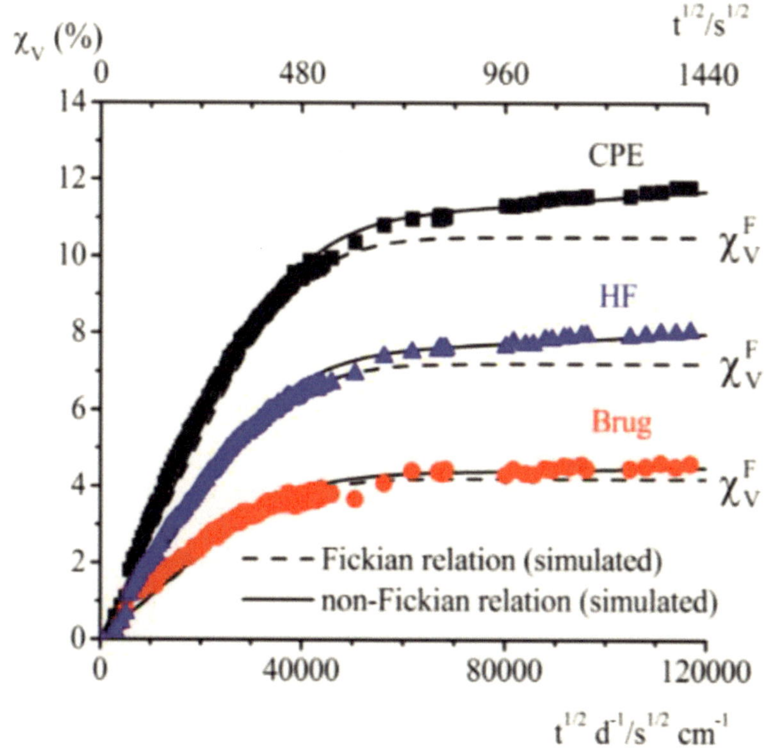

Figure 14. Evolution of the volumetric water content χ_v for unstressed coating films in a 3 wt.% NaCl solution, immersed at 30°C, calculated using the film capacitance at high frequencies C_{HF}, the film capacitance equivalent CPE, and the "true" film capacitance C_{Brug} [68].

4. Discussion

In the previous section, the most common techniques to measure or follow the water uptake in organic coatings were introduced, complemented with the time-resolved or spectrally resolved THz-method that was recently developed. **Table 1** shows the experimental characteristics and the limitations and **Table 2** shows the advantages and disadvantages of each technique.

In most diffusion studies, researchers work on two or more techniques to obtain complementary information. No matter the number of studies done so far to extract the diffusion coefficients of the electrolyte, the detailed analysis of moisture diffusion in polymer film and coatings—including the dependence on moisture content, temperature, salt concentration, pigments,…—has never been comprehensively determined. In fact, the determination of such relationships requires the measuring of time-dependent penetrant concentration profiles in thin polymer film or coatings. This is hard to achieve, not only because of the heterogeneous nature of most coatings, but also because of the insufficient penetration depth (ATR-FTIR) or depth resolution (NMR, terahertz spectroscopy), or because of the inability of the measuring techniques to study the concentration gradient directly (GM, EIS). THz spectroscopy (THz-TDS and THz-FDS), EIS, and GM can be applied to coatings of thicknesses of several millimeters. For thick coatings, the high impedance limits the suitability of commercial instrumentation.

	GM	ATR-FTIR	NMR	THz: TDS and FDS	EIS
In situ/ex situ	Ex situ	In situ and ex situ	In situ and ex situ	In situ and ex situ	In situ
Penetration depth d_p	$d_p \geq \mu m$	$nm \leq d_p \leq 5\ \mu m$	$d_p \geq \mu m$	$\mu m \leq d_p \leq cm$	$\mu m \leq d_p \leq mm$
Measurements	Weight difference between wet/dry conditions	Peak intensities corresponding to OH absorptions	Change in signal intensity depth profiles between wet and dry coatings	Imaging: contrast between wet and dry regions Coatings: change in reflection coefficient	Coating capacitance derived from impedance spectra using electrical equivalent models

Table 1. Experimental characteristics of all techniques.

The more complex the coating is, the more difficult it is to obtain valid diffusion coefficients. Methodological difficulties of separating the contributions of adsorption, absorption, capillary condensation…, make it difficult to establish a single model for obtaining accurate quantitative evaluations of an organic coating. In general, these effects are studied qualitatively rather than quantitatively. Additives, pigments and fillers further affect the equilibrium parameters (solubility or equilibrium water volume fraction) and the diffusion kinetic parameters (diffusion coefficient and permeability). For example, the presence of additives in the polymer can significantly increase solubility and equilibrium sorption values of water in the coating, thereby increasing the permeability. Of the techniques considered, EIS is most sensitive to (macroscopic) heterogeneity and defects in the coating (pinholes, particle aggregates,…).

For some of the techniques (ATR-FTIR, NMR) the use of free-standing films can be advantageous, however, this may affect the water migration. Holtzman [70] showed that transport of water in coatings applied on substrates is slightly lower than in free films. However, Rosen and Martin [71] found that in epoxy coatings applied to steel substrates, the diffusion coefficient is three times greater than those in free films. Internal stresses residing in the polymer after production probably play an important role. So, care must be taken in directly comparing results of layers prepared in different ways (free-standing or not, substrates, curing or drying protocol,…).

GM is a standard technique that allows fast measurements of the weight difference between the dry and the wet sample. The main drawback of this method is its sensitivity to the sample handling: a small excess of liquid on the surface or evaporation losses of volatile components may have a big impact on the results. Therefore, this method is not suitable for studying the initial stages of diffusion. However, it is often used as a complementary technique.

EIS is a powerful technique as it gives information on the very early stages of the water uptake process, in addition to information on other processes, such as the subsequent corrosion process. However, the data are interpreted based on equivalent electric circuits. Hence, the validity of the result depends on the validity of the chosen model.

NMR relaxometry using the GARField approach and THz-TDS and THz-FDS look quite powerful but are not generally accessible yet.

	GM
Pros	• Direct and fast measurement
	• Simple experimental set-up
	• Low cost equipment
	• Often used as complementary technique
Cons	• Discontinuous measurements
	• Risk of low precision due to fast evaporation before measurement
	• Risk of low precision due to water surface layers of variable thickness
	• For small amounts of water absorbed in coatings on metal substrates: very small relative mass changes
	ATR-FTIR
Pros	• Simple experimental set-up
	• Information on interaction between water and polymer molecules
	• Can probe distribution of water parallel to the coating/substrate interface
	• On free-standing films, it is feasible to follow in situ and continuously the migration of water
	• Good reproducibility
Cons	• Small penetration depth
	• On coatings: discontinuous measurements
	• Contact between sample and ATR crystal is crucial and may change as coating absorbs water
	NMR relaxometry (GARField)
Pros	• Gradients in moisture content in 100-μm coatings on glass substrates can be measured with 3–6 μm depth resolution
	• Probe distribution of water in multilayered coatings with multiphase polymer layers
	• Distinction between free and bound water, interaction between water and polymer molecules
Cons	• Metal substrates interfere with NMR, but taking this into account might be feasible [68]
	• Probe not yet commercially available
	THz: TDS & FDS
Pros	• Very fast measurement
	• Feasible on coated metal
	• High penetration depth up to a few cm for the lowest frequencies of the sub-THz spectrum
	• In thicker coatings (>100 μm), measuring gradients might be feasible
	• Imaging mode: water diffusion into comparably-sized cracks, voids
	• Advanced frequency domain techniques allow very high resolution in the water content measurements, down to 0.0005%
	• Transient radar methodology allows simultaneous measurements of (swelling) thickness of the coating and its water content via a dielectric mixing model
Cons	• Application on water diffusion in coatings was introduced recently (limited set of results exists)
	• Basic instrument yet commercially available
	• More advanced sensor systems need to be custom designed
	EIS
Pros	• Fast measurements at high frequencies
	• Normally on coated metal, but feasible on free-standing films
	• Complete view of coating deterioration process. Distinction of all phenomena that occur in the system (water ingress, double layer creation at the interface, beginning of corrosion)
	• Measuring the decrease of the coating resistance gives qualitative and quantitative information about the coating performance
Cons	• Slow measurements at low frequencies
	• If the potential perturbation is too high, it may affect the coated/metal system (non-linear behavior)
	• Interpretation of measurement data based on equivalent electrical circuits
	• Results are sensitive to pin holes and coating defects in general
	• Surface treatment of metal substrate before coating application may affect the measurements

Table 2. Pros and cons of all techniques for measuring diffusion of water in coatings.

5. Conclusion

Organic coatings are widely used for corrosion protection of metallic structures and the detailed characterization of water ingress in coatings is therefore crucial. The complexity of organic coatings (chemical composition, crosslinking), the presence of a wide range of additives for ensuring a wide range of properties, in combination with the nature of the metal substrate and its surface treatments, makes the research of water uptake in coatings quite challenging. Each of the techniques presented above has advantages and limitations, so selecting the most appropriate one strongly depends on the purposes of the research and the coating characteristics. Knowing the diffusion coefficient and how it is affected by the different factors (composition, temperature, filler, aging…) is crucial for making reliable life time predictions of coatings based on limited sets of experimental results. Often, to illustrate the capability of a technique, model coatings are chosen in accordance with the limitations of that technique. Of course, to measure the diffusion behavior, the coatings must fulfill some criteria, such as chemical stability, homogeneous composition, well-dispersed fillers, defect free,…, otherwise the results will never be reliable. The basic characteristics and limitations of each technique are pointed out in **Table 1**, while the main advantages and disadvantages are discussed in **Table 2**. Given these points, it is clear that a combination of two or more techniques will usually be needed to provide a clearer overview of the water diffusion process in organic coatings.

Conflict of interest

The authors whose names are listed immediately above certify that they have no affiliations with or involvement in any organization or entity with any financial interest, or non-financial interest in the subject matter or materials discussed in this manuscript.

Author details

Zoi Manoli[1]*, Darja Pecko[1], Guy Van Assche[2], Johan Stiens[3], Ali Pourkazemi[3] and Herman Terryn[1]

*Address all correspondence to: zoi.manoli@vub.be

1 Research Group of Electrochemical and Surface Engineering (SURF), Vrije Universiteit Brussel (VUB), Brussels, Belgium

2 Physical Chemistry and Polymer Science (FYSC), Vrije Universiteit Brussel (VUB), Brussels, Belgium

3 Department of Electronics and Informatics (ETRO), Vrije Universiteit Brussel (VUB), Brussels, Belgium

References

[1] Funke W. Toward a unified view of the mechanism responsible for paint defects by metallic corrosion. Industrial & Engineering Chemistry Product Research and Development. 1985;**24**:343-347. DOI: 10.1021/i300019a001

[2] Stratmann M, Fese R, Leng A. Corrosion protection by organic films. Electrochimica Acta. 1994;**39**:1207-1214. DOI: 10.1016/0013-4686(94)E0038-2

[3] García SJ, Fischer HR, van der Zwaag S. A critical appraisal of the potential of self healing polymeric coatings. Progress in Organic Coating. 2011;**72**:211-221

[4] Bergman SD, Wudl F. Mendable polymers. Journal of Materials Chemistry. 2008;**18**:41-62. DOI: 10.1016/j.porgcoat.2011.06.016

[5] Szauer T. Electrical and electrochemical resistances for the evaluation of protective nonmetallic coatings. Progress in Organic Coating. 1982;**10**:157-170. DOI: 10.1016/0300-9440(82)80014-3

[6] Inone PC, Garcia CM, Ruvolo-Filho A. Evaluating barrier properties of organic coatings by water permeation and electrochemical methods. Journal of Coatings Technology. 2003;**75**:29-36. DOI: 10.1007/BF02697918

[7] Henderson MA. The interaction of water with solid surfaces fundamental aspects revisited. Surface Science Reports. 2002;**46**:1-308. DOI: 10.1016/S0167-5729(01)00020-6

[8] Posner R, Ozcan O, Grundmeier G. Water and ions at polymer/metal interfaces. Posner. In: da Silva LFM, Sato C, editors. Design of Adhesive Joints under Humid Conditions. Vol. 25. Berlin: Springer; 2013. pp. 21-52. DOI: 10.1007/978-3-642-37614-6_2

[9] van Ooij WJ, Sabata A, Loison D, Jossic T, Charbonnier JC, Adhes J. Paint delamination from electrocoated automotive steels during atmospheric corrosion. Part I. Hot-dip galvanized and electrogalvanized steel. Science and Technology. 1989;**3**:1-27. DOI: 10.1163/156856189X00010

[10] Ogle K, Morel S, Meddahi N. An electrochemical study of the delamination of polymer coatings on galvanized steel. Corrosion Science. 2005;**47**:2034-2052. DOI: 10.016/j.corsci.2004.08.017

[11] Leng A, Streckel H, Stratmann M. The delamination of polymeric coatings from steel. Part 1: Calibration of the Kelvinprobe and basic delamination mechanism. Corrosion Science. 1999;**41**:574-557. DOI: 10.1016/S0010-938X(98)00166-8

[12] Bellenger V, Verdu J, Morel E. Structure-properties relationships for densely cross-linked epoxide-amine systems based on epoxide or amine mixtures. Journal of Materials Science. 1989;**24**:63-68. DOI: 10.1007/BF00660933

[13] Apicella A, Nicolais L, de Cataldis C. Characterization of the morphological fine structure of commercial thermosetting resins through hygrothermal experiments. In: Kaush HH, Zachman HG, editors. Characterization of Polymers in the Solid State I: Part A: NMR and

Other Spectroscopic Methods Part B: Mechanical Methods. Vol. 66. Berlin, Heidelberg: Springer; 1985. pp. 189-207. DOI: 10.1007/3-540-13779-3_21

[14] Adamson MJ. Thermal expansion and swelling of cured epoxy resin used in graphite/epoxy composite materials. Journal of Applied Physics. 1980;**15**:1736-1745. DOI: DOI 10.1007/BF00550593

[15] Apicella A, Tessieri R, De Cataldis C. Sorption modes of water in glassy epoxies. Journal of Membrane Science. 1984;**18**:211-225. DOI: 10.1016/S0376-7388(00)85035-8

[16] Johncock P, Tudgey GF. Some effects of structure, composition and cure on the water absorption and glass transition temperature of amine-cured epoxies. British Polymer Journa. 1986;**18**:292-302. DOI: 10.1002/pi.4980180504

[17] Fujita H, Kishimoto A, Matsumoto K. Concentration and temperature dependence of diffusion coefficients for systems polymethyl acrylate and n-alkyl acetates. Transactions of the Faraday Society. 1960;**56**:424-437. DOI: 10.1039/TF9605600424

[18] Prager S, Bagley E, Long FA. Diffusion of hydrocarbon vapors into polyiso-butylene. II. Journal of the American Chemical Society. 1953;**75**:1255-1256. DOI: 10.1021/ja01101a517

[19] Soney CG, Sabu T. Transport phenomena through polymeric systems. Progress in Polymer Science. 2001;**26**:985-1017. DOI: 10.1016/S0079-6700(00)00036-8

[20] Zhou J, Lucas JP. Hygrothermal effects of epoxy resin. Part I: The nature of water in epoxy. Polymer. 1999;**40**:5505-5512. DOI: 10.1016/S0032-3861(98)00790-3

[21] Zhou J, Lucas JP. Hygrothermal effects of epoxy resin. Part II: Variations of glass transition temperature. Polymer. 1999;**40**:5513-5522. DOI: 10.1016/S0032-3861(98)00791-5

[22] Bouvet G, Dang N, Cohendoz S, Feaugas X, Mallarino S, Touzain S. Impact of polar groups concentration and free volume on water sorption in model epoxy free films and coatings. Progress in Organic Coating. 2016;**96**:32-41. DOI: 10.1016/j.porgcoat.2015.12.011

[23] Lyon S, Bingham R, Mills DJ. Advances in corrosion protection by organic coatings: What we know and what we'd like to know. Progress in Organic Coating. 2017;**102**(Part A):2-7. DOI: 10.1016/j.porgcoat.2016.04.030

[24] Crank J, Park GS. Methods of measurement. In: Crank J, Park GS, editors. Diffusion in Polymers. London, UK: Academic Press, Inc.; 1968. pp. 1-39

[25] Kloppers MJ, Bellucci F, Latanision RM, Brennan JE. Transport of dielectric properties of poly(ethylene terephthalate) as determined via electrochemical techniques. Journal of Applied Polymer Science. 1993;**48**:2063-2252. DOI: 10.1002/app.1993.070481213

[26] Fieldson GT, Barbari TA. The use of FTIR-ATR spectroscopy to characterize penetrant diffusion in polymers. Polymer. 1993;**34**:1146-1156. DOI: 10.1016/0032-3861(93)90765-3

[27] Fieldson GT, Barbari TA. Analysis of diffusion in polymers using evanescent field spectroscopy. AICHE Journal. 1995;**41**:795-804. DOI: 10.1002/aic.690410406

[28] Nguyen T, Byrd E, Lin C. A spectroscopic technique for in situ measurement of water at the coating/metal interface. Journal of Adhesion Science and Technology. 1991;**5**:697-709. DOI: 10.1163/156856191X00648

[29] Philippe L, Lyon S, Sammon C, Yarwood J. An FTIR/ATR in situ study of sorption and transport in corrosion protective organic coatings: 1. Water sorption and the role of inhibitor anions. Progress in Organic Coating. 2004;**49**:302-314. DOI: 10.1016/j.porgcoat.2003.07.002

[30] Pereira MR, Yarwood J. ATR-FTIR spectroscopic studies of the structure and permeability of sulfonated poly(ether sulfone) membranes. Part 1. Interfacial water-polymer interactions. Journal of the Chemical Society, Faraday Transactions. 1996;**92**:2731-2735. DOI: 10.1039/FT9969202731

[31] Pereira MR, Yarwood J. ATR-FTIR spectroscopic studies of the structure and permeability of sulfonated poly(ether sulfone) membranes. Part 2. Water diffusion processes. Journal of the Chemical Society, Faraday Transactions. 1996;**93**:2737-2743. DOI: 10.1039/FT9969202737

[32] Hajatdoost S, Yarwood J. ATR-FTIR spectroscopic studies of the structure and permeability of sulfonated poly(ether sulfone) membranes. Part 3. Effects of sorption and desorption, and of annealing. Journal of the Chemical Society, Faraday Transactions. 1997;**93**: 1613-1620. DOI: 10.1039/A608447H

[33] Yarwood J, Sammon C, Mura C, Pereira MR. Vibrational spectroscopic studies of the diffusion and perturbation of water in polymeric membranes. Journal of Molecular Liquids. 1999;**80**:93-115. DOI: 10.1016/S0167-7322(99)80002-6

[34] Sammon C, Yarwood J, Everall N. A FTIR–ATR study of liquid diffusion processes in PET films: Comparison of water with simple alcohols. Polymer. 2000;**41**:2521-2534. DOI: 10.1016/S0032-3861(99)00405-X

[35] Wapner K, Stratmann M, Grundmeier G. In situ infrared spectroscopic and scanning Kelvin probe measurements of water and ion transport at polymer/metal interfaces. Electrochimica Acta. 2006;**51**:3303-3315. DOI: 10.1016/j.electacta.2005.09.024

[36] Wapner K, Grundmeier G. Spatially resolved measurements of the diffusion of water in a model adhesive/silicon lap joint using FTIR-transmission-microscopy. International Journal of Adhesion and Adhesives. 2004;**24**:193-200. DOI: 10.1016/j.ijadhadh.2003.09.008

[37] Nguyen T, Byrd E, Bentz D, Lin C. In situ measurement of water at the organic coating/substrate interface. Progress in Organic Coating. 1996;**27**:181-193. DOI: 10.1016/0300-9440(95)00535-8

[38] Baukh V, Huinink HP, Adan OCG, Sebastiaan J, Erich F, van der Ven LGJ. NMR imaging of water uptake in multilayer polymeric films: Stressing the role of mechanical stress. Macromolecules. 2010;**43**:3882-3889. DOI: 10.1021/ma1001996

[39] Bauer DR. Characterization of crosslinked polymers by high-resolution solids nuclear magnetic resonance spectroscopy. Progress in Organic Coating. 1986;**14**:45-65. DOI: 10.1016/0033-0655(86)80015-2

[40] Carr HY, Purcell EM. Effects of diffusion on free precession in nuclear magnetic resonance experiments. Physics Review. 1954;**94**:630-638. DOI: 10.1103/PhysRev.94.630

[41] Kimmich R, Fatkullin N. Polymer chain dynamics and NMR. Advances in Polymer Science. 2004;**170**:1-113. DOI: 10.1007/978-3-540-40000-4_5

[42] Adriaensens P, Pollaris A, Carleer R, Vanderzande D, Gelan J, Litvinov VM, et al. Quantitative magnetic resonance imaging study of water uptake by polyamide 4,6 (PA46) plates. Polymer. 2001;**42**:7943-7952. DOI: 10.1016/S0032-3861(01)00314-7

[43] Brownstein KR, Tarr CE. Importance of classical diffusion in NMR studies of water in biological cells. Physical Review A. 1979;**19**:2446-2453. DOI: 10.1103/PhysRevA.19.2446

[44] Araujo CD, Mackay AL, Whittall KP, Hailey JRT. A diffusion model for spin-spin relaxation of compartmentalized water in wood. Journal of Magnetic Resonance, Series B. 1993; **101**:248-261. DOI: 10.1006/jmrb.1993.1041

[45] Glover PM, Aptaker PS, Bowler JR, Ciampi E, McDonald PJ. A novel high-gradient permanent magnet for the profiling of planar films and coatings. Journal of Magnetic Resonance. 1999;**139**:90-97. DOI: 10.1006/jmre.1999.1772

[46] Aptaker PS, McDonald PJ, Mitchell J. Surface GARField: A novel one-sided NMR magnet and RF probe. Magnetic Resonance Imaging. 2007;**25**:548. DOI: 10.1016/j.mri.2007.01.020

[47] Zhu H, Huinink HP, Adan OCG, Kopinga K. NMR study of the microstructures and water-polymer interactions in cross-linked polyurethane coatings. Macromolecules. 2013; **46**:6124-6131. DOI: 10.1021/ma401256n

[48] Baukh V, Huinink HP, Adan OCG, Erich SJF, van der Ven LGJ. Water-polymer interactions during water. Macromolecules. 2011;**44**:4863-4871. DOI: 10.1021/ma102889u

[49] Federici JF. Review of moisture and liquid detection and mapping using terahertz imaging. Journal of Infrared, Millimeter and Terahertz Waves. 2012;**33**:97-126. DOI: 10.1007/s10762-011-9865-7

[50] Hu BB, Nuss MC. Imaging with terahertz waves. Optics Letters. 1995;**20**:1716-1718. DOI: 10.1364/OL.20.001716

[51] Mittleman DM, Jacobsen RH, Nuss MC. T-Ray imaging. IEEE Journal of Selected Topics in Quantum Electronics. 1996;**2**:679-692. DOI: 10.1109/2944.571768

[52] Jordens C, Wietzke S, Scheller M, Koch M. Investigation of the water absorption in polyamide and wood plastic composite by terahertz time-domain spectroscopy. Polymer Testing. 2010;**29**:209-2015. DOI: 10.1016/j.polymertesting.2009.04.010

[53] Stiens J et al. Enhanced Characterization of Dielectric Materials, EP15189446. 2015

[54] Pandey G. Ultra-sensitive millimeter wave sensor for material characterization [thesis]. Brussels, Belgium: Vrije Universiteit Brussel; 2017

[55] Pandey G, Vandermeiren W, Dimiccoli L, Stiens J. Contactless monitoring of food drying and freezing processes with millimeter waves. Journal of Food Engineering. 2018;**226**:1-8. DOI: 10.1016/j.jfoodeng.2018.01.003

[56] Pourkazemi A, Stiens JH, Becquaert M, Vanderwal M. Transient radar method: Novel illumination and blind electromagnetic/geometrical parameter extraction technique for multilayer structures. IEEE Transactions on Microwave Theory and Techniques. 2017;**65**: 2171-2184. DOI: 10.1109/TMTT.2017.2665633

[57] Obradovic J, Collins JHP, Hirsch O, Mantle MD, Johns ML, Gladden LF. The use of THz time-domain reflection measurements to investigate solvent diffusion in polymers. Polymer (Guildf). 2007;**48**:3494-3503. DOI: 10.1016/j.polymer.2007.04.010

[58] Kusano M, Kubouchi M, Bulgarevich DS, Shiwa M. Non-destructive evaluation by terahertz spectroscopy for penetration of acid solutions into epoxy resin. Express Polymer Letters. 2016;**10**:941-949. DOI: 10.3144/expresspolymlett.2016.87

[59] Deflorian F, Rossi S. An EIS study of ion diffusion through organic coatings. Electrochimica Acta. 2006;**51**:1736-1744. DOI: 10.1016/j.electacta.2005.02.145

[60] Castela ASL, Simões AM, Ferreira MGS. EIS evaluation of attached and free polymer films. Progress in Organic Coating. 2000;**38**:1-7. DOI: 10.1016/S0300-9440(99)00076-4

[61] Fredj N, Cohendoz S, Mallarino S, Feaugas X, Touzain S. Evidencing antagonist effects of water uptake and leaching processes in marine organic coatings by gravimetry and EIS. Progress in Organic Coating. 2010;**67**:287-295. DOI: 10.1016/j.porgcoat.2009.11.001

[62] Deflorian F, Fedrizzi L, Rossi S, Bonora PL. Organic coating capacitance measurement by EIS: Ideal and actual trends. Electrochimica Acta. 1999;**44**:4243-4249. DOI: 10.1016/S0013-4686(99)00139-5

[63] Brasher DM, Kingsbury AH. Electrical measurements in the study of immersed paint coatings on metal. I. Comparison between capacitance and gravimetric methods of estimating water-uptake. Journal of Applied Chemistry. 1954;**4**:62-72. DOI: 10.1002/jctb.5010040202

[64] Lindqvist SA. Theory of dielectric properties of heterogeneous substances applied to water in a paint film. Corrosion. 1985;**41**(2):69-75. DOI: 10.5006/1.3581974

[65] Galliano F, Landolt D. Evaluation of corrosion protection properties of additives for water-borne epoxy coatings on steel. Progress in Organic Coating. 2002;**44**:217-225. DOI: 10.1016/S0300-9440(02)00016-4

[66] Deflorian F, Fedrizzi L, Bonora PL. Influence of the photo-oxidative degradation on the water barrier and corrosion protection properties of polyester paints. Corrosion Science. 1996;**38**:1697-1708. DOI: 10.1016/S0010-938X(96)00062-5

[67] Bellucci F, Nicodemo L. Water transport in organic coatings. Corrosion. 1993;**49**:235-247. DOI: 10.5006/1.3316044

[68] Nguyen Dang D, Peraudeau B, Cohendoz S, Mallarino S, Feaugas X, Touzain S. Effect of mechanical stresses on epoxy coating ageing approached by electrochemical impedance spectroscopy measurements. Electrochimica Acta. 2014;**124**:80-89. DOI: 10.1016/j.electacta.2013.08.111

[69] Zhu H, Huinink HP, Erich SJ, Baukh V, Adan OC, Kopinga K. High spatial resolution NMR imaging of polymer layers on metallic substrates. Journal of Magnetic Resonance. 2011;**214**:227-236. DOI: 10.1016/j.jmr.2011.11.009

[70] Holtzman KA. Water vapor transport in adherent organic coatings. Journal of Paint Technology. 1971;**554**:47

[71] Rosen HN, Martin JM. Sorption of moisture on epoxy and alkyd free films and coated steel panels. Journal of Coatings Technology and Research. 1991;**63**:85

Chapter 7

In Situ Renewable Coating of Boron Carbide (B_4C) for Plasma Materials for Plasma-Technological and Fusion Devices

Leon Begrambekov, Andrey Grunin,
Nikolay Puntakov, Yaroslav Sadovskiy,
Vyacheslav Budaev and Sergei Grashin

Additional information is available at the end of the chapter

http://dx.doi.org/10.5772/intechopen.81361

Abstract

The application of the in situ renewable protecting boron carbide (B_4C) coating can prevent plasma-facing materials of plasma technology and thermonuclear devices from plasma irradiation and by this means prevents their destruction and plasma contamination by materials of their erosion. At the same time, the regimes and conditions of high adhesive deposition of B_4C on tungsten and the B_4C coating ability to withstand the thermal cycling and high-power density irradiation by plasma ions have not been investigated yet. The chapter considers the results of ion irradiation and thermal cycling of boron carbide coating on tungsten sample in Stand for Coating Deposition and Material Testing—CODMATT (NRNU MEPhI) and plasma irradiation during a plasma disruption in fusion device—T-10 tokamak (NRC "Kurchatov Institute"). Boron carbide coating withstands the thermal cycling and high-power density irradiation by plasma ions. It retains uniformity and adhesion to tungsten and protects it from direct plasma interaction for temperatures up to melting point of tungsten. The retaining of uniform coating in contact with tungsten substrate allows renewing the coating on its surface even after high-energy plasma loads.

Keywords: investigation, boron carbide, coating, ion cycling, plasma

1. Introduction

Tungsten is used as the plasma-facing material of the contemporary fusion devices and for the divertor of the International Thermonuclear Experimental Reactor (ITER) under construction.

Last year investigations show that high-power density plasma irradiation initiates cracking of tungsten surface, blister formation, flaking, macroscopic particle emission, etc. (see, for instance, [1, 2]). These phenomena can cause accelerated destruction of tungsten tiles of ITER divertor.

It was shown [3] that application of the in situ renewable protecting boron carbide (B_4C) coating can prevent plasma-facing materials of plasma technology and thermonuclear devices from plasma irradiation and by this means prevents their destruction and plasma contamination by materials of their erosion. B_4C coating has high melting temperature (2800 K). The erosion rate of B_4C changes only slightly up to 1400°C and appears to be much less than the erosion rate of dense graphites under similar irradiation conditions [4]. The B_4C coating being used as the protective coating of the plasma-facing components of fusion devices will not have thickness higher than 30–50 μm; therefore, accumulation of remarkable amount of tritium in the plasma-facing materials is prevented.

Investigations showed that near stoichiometric B_4C coating can be deposited in high-temperature plasma providing total dissociation of the molecules of initial substance. The conclusion was confirmed experimentally, when B_4C coating had been successfully deposited in plasma devices (PISCEC-B [5] and tokamak T11-M [6]). The vapor of nontoxic, nonexplosive, and nonhazardous carborane ($C_2B_{10}H_{12}$) was used as the initial substance for coating deposition in the course of regular discharge in the plasma devices.

At the same time, some of the important aspects of the B_4C coating application in fusion devices have not been investigated yet. Among them, there are the regimes and conditions of high adhesive deposition of B_4C on tungsten and the B_4C coating ability to withstand the thermal cycling and high-power density irradiation by plasma ions.

The results of these topics' investigations are the subject of this presentation. The paper considers the results of ion irradiation and thermal cycling of boron carbide coating on tungsten sample in Stand for Coating Deposition and Material Testing, CODMATT (National Research Nuclear University (MEPhI)) [7], and plasma irradiation during a plasma disruption in fusion device (T-10 tokamak, National Research Center "Kurchatov Institute," Moscow).

2. Boron carbide deposition and testing in the laboratory stand

The method of B_4C coating deposition on tungsten, reproducing in general terms the coating regime in conditions of fusion devices, was developed on the CODMATT facility. The coating was formed by atoms of boron and carbon sputtered by plasma ions from corresponding targets. For tests in T-10, a coating of boron carbide was deposited on tungsten samples measuring 15 × 15 × 1 mm^3, 5 μm in thickness. The photograph of the coating is shown in **Figure 1**.

It is seen that coatings had a smooth surface, cracks, and peeling and other signs of a violation of adhesion to the substrate were not observed. Using X-ray energy dispersive spectrometry (EDS), it was found that the coating practically corresponds to the stoichiometric composition of boron carbide (B76%, C22%).

Several cycles of pulsed irradiation of tungsten samples coated with hydrogen ions on the CODMATT facility were carried out. In the first cycle, the irradiation flux power density was

http://dx.doi.org/10.5772/intechopen.81361

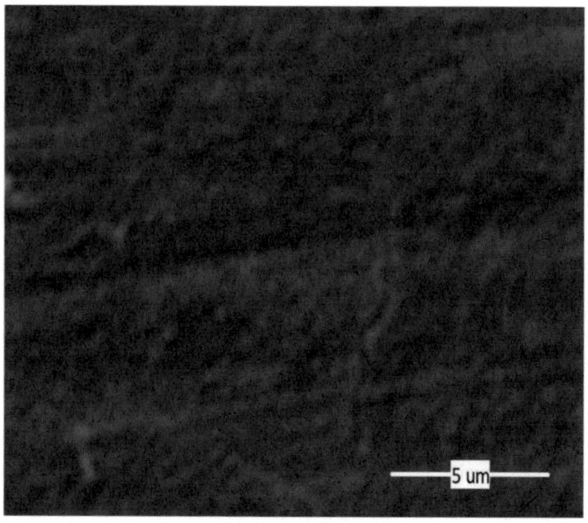

Figure 1. Boron carbide coating on tungsten substrate.

0.3 MW/m^2. The energy of the irradiating ions $E_i = 10$ kV/at, the duration of irradiation $\tau = 0.6$ s, the total cycle time t = 300 s, and the temperature of the sample during the test varied in the range T = 100–400°C. One thousand pulses were conducted. The second cycle was performed using irradiation flux power density which equals 5 MW/m^2 ($E_i = 15$ kV/at, W = 5.0 MW/m^2, $\tau = 0.4$ s, t = 30 s, T = 600–900°C). One hundred pulses were conducted. During the third cycle irradiation, flux power density was elevated up to 5 MW/m^2 ($E_i = 15$ kV/at, W = 5.0 MW/m^2, $\tau = 0.6$ s, t = 30 s, T = 700–1200°C). Fifty pulses were conducted.

None of the test cycles led to the peeling of the coatings, the appearance of caverns, cracks, and changes in the composition of the coating. Noticeable traces of etching of the surface appeared only after the last cycle.

3. T-10 boron carbide coating tests in a T-10 tokamak

A tungsten sample with a coating located on the diagnostic input of the T-10 tokamak was exposed to plasma irradiation during a plasma disruption.

Part of the sample subjected to the most intense plasma irradiation melted out and left the sample (**Figure 2**). Surface morphology analysis of the remaining part of the sample after the irradiation was done on a TESCAN VEGA 3 scanning electron microscope (SEM). Five areas are distinctly visible on a sample around the molten part: the area of tungsten that melted but stayed from the sample; the area of melted coating, the upper level of which collected in separate globules; the area of partial coating melting; the area of cracked non-molten coating; and the area covered by the sample holder.

An SEM image of the molten tungsten area of the sample is shown in **Figure 3**. Distinct micro-crystals are visible which were formed due to solidification of the molten tungsten, as well as pores and trenches on the tungsten grain boundaries. Measurements performed by EDS have not shown the presence of boron and carbon in a layer about 0.8 μm thick, which indicates complete or almost complete removal of coating from the surface of molten tungsten.

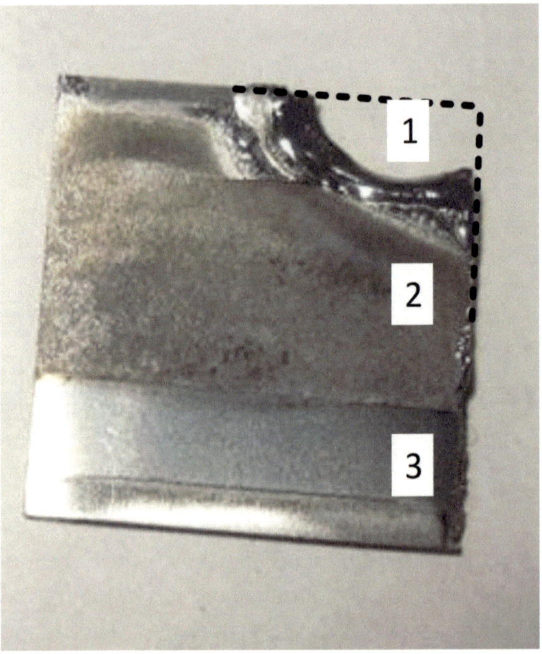

Figure 2. Sample after tests: (1) tungsten melting area, (2) irradiated part of the coating, and (3) sample holder area.

Area of a molten coating is shown in **Figure 4**. Multiple globules with sizes of tens of micrometers are of notice on this area, which were formed, supposedly, under surface tension force on the upper part of the coating. The coating left on the surface of the sample is a uniform layer with small cracks, but no signs of exfoliation.

For measuring the thickness of coating left on a sample, a cross section was performed using a FEI Scios DualBeam device with a focused ion beam method. The cross section went through one of the cracks on the area free of globules (**Figure 5**).

A substrate, an intermediate layer, the remaining part of the coating (approx. 1.8 μm), and a crack can be seen on the cross section of the sample. The crack is visible in the tungsten substrate, which indicates that this and, supposedly, other cracks were initiated in the substrate, with cracks forming during the cooling of tungsten at temperatures lower than boron carbide freezing point and going on the crystal grain boundaries, as seen in **Figure 5**. An SEM image of coating remaining on the sample, done in a large magnification (**Figure 6**), shows its porosity, with pores that probably form during the solidification of the molten coating.

Globules and coating remaining on the surface of the substrate, according to EDS analysis, included approximately identical parts of boron and carbon. Their ratios ranged from 1:1 to 1.2:1 boron to carbon; however, it is known that boron carbide structure remains even at a ratio of B:C = 1:1. Some percentage in globule and remaining coating content was stainless steel components (Fe, Ni, Cr) and oxygen, which were probably the tokamak's working gas impurities introduced into the coating during plasma irradiation.

The area of a partially molten coating is shown in **Figure 7**. A number of characteristic areas can be observed on this part, such as initial coating with cracks, molten parts, and points of

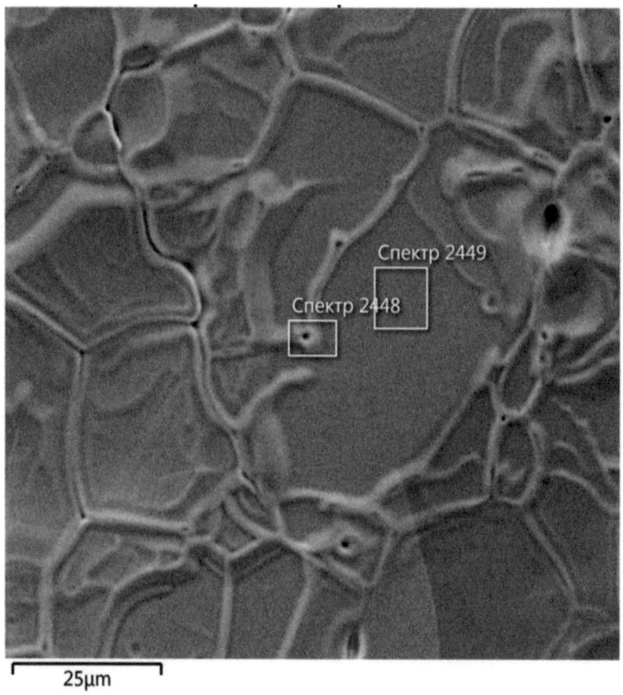

Figure 3. Molten tungsten area of a sample.

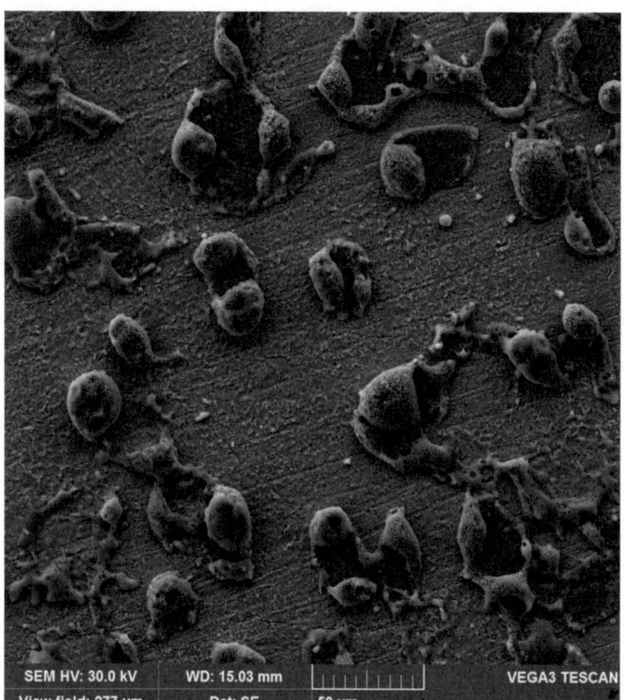

Figure 4. Area of a molten coating and globules.

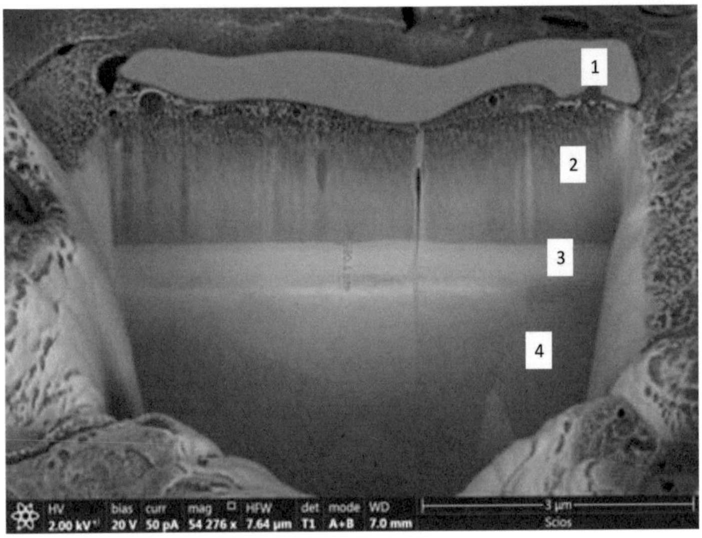

Figure 5. Cross section of a globule-free sample area. (1) Pt technical layer, (2) coating, (3) intermediate layer, and (4) substrate.

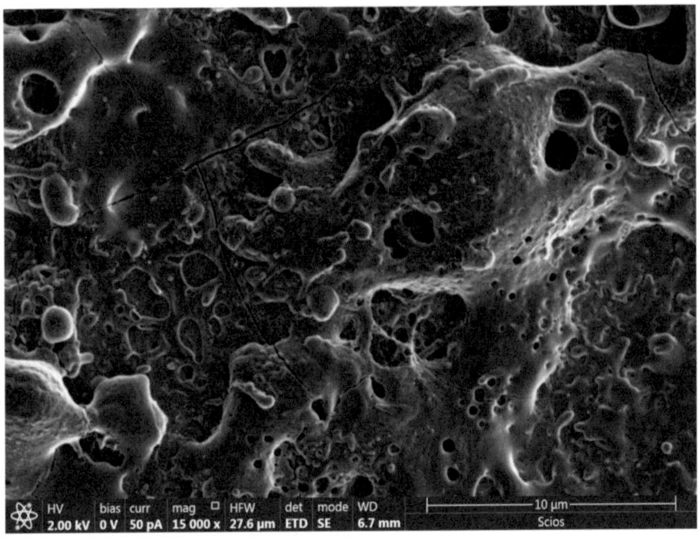

Figure 6. Area with a crack on the remaining coating.

molten coating accumulation, with the latter not collecting into globules, but rather collecting on the peripherals of molten areas. The size of molten parts ranges from 5 to 200 μm. The ratios of boron to carbon in these areas were 2.3:1, 2.6:1 and 3.1:1, respectively. Thus, it can be stated that the content of the coating on this area of the sample was much closer to the original stoichiometry than on the previous one.

The last characteristic area on a sample is shown in **Figure 8**.

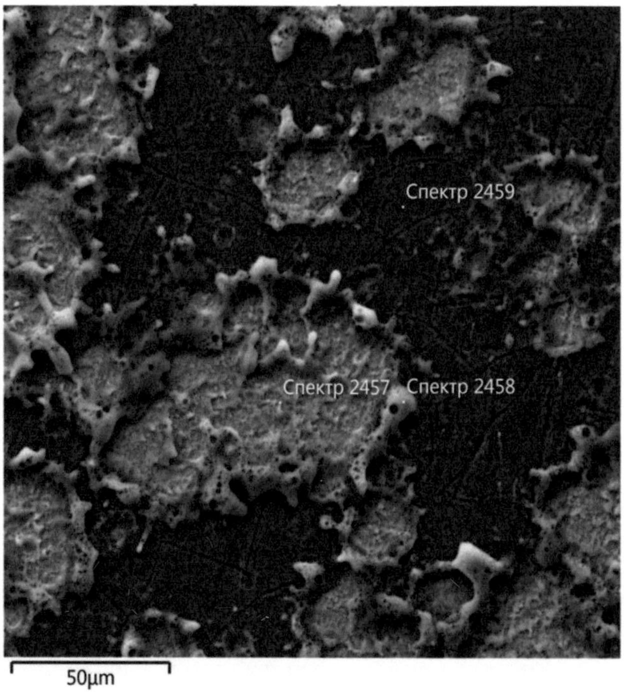

Figure 7. Area of a partially molten coating.

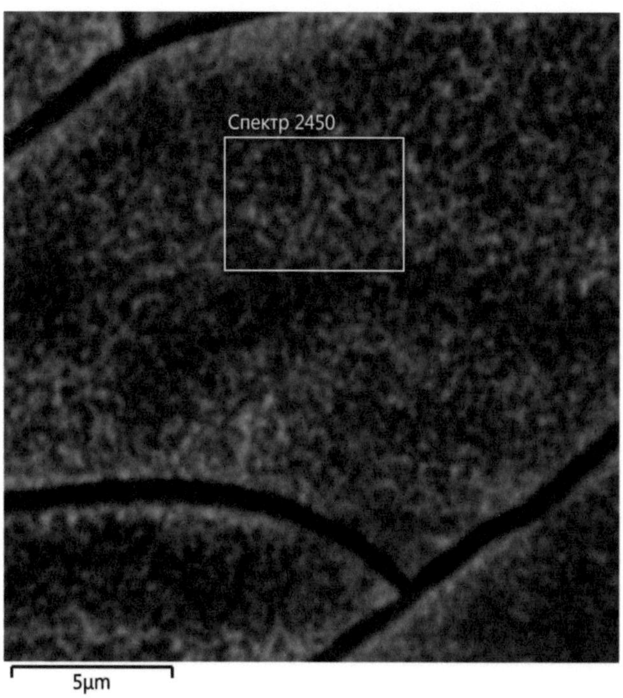

Figure 8. Area of nonirradiated coating with cracks.

There was no trace of coating melting due to plasma interaction. There was also no indication of change in the coating thickness, and EDS analysis did not show any change in element content. The appearance of cracks shows that destruction of tungsten occurred even on this part of substrate, which was positioned next to a holder and heated less than the rest of the sample.

4. Measurement of gas trapped in the sample

Analysis of ion irradiation parameters during plasma disruption has shown that the coating was irradiated by ions with an average energy of approx. 100 eV with a mean dose of approx. 1×10^{18} at/cm^2. Measurement of gas content trapped in the coating during plasma irradiation was done using a method of thermal desorption spectroscopy (TDS). The analysis was performed for two parts of the samples: a partially molten area (**Figure 7**) and an intact area (**Figure 8**). TDS analysis of both these parts is shown in **Figure 9**. The amount of trapped deuterium is 9×10^{14} at./cm^2 for the area with a partially molten coating and 2×10^{14} at./cm^2 for the area with cracking.

The temperature of the maxims and high-temperature edge of the spectrum of deuterium thermal desorption from the coating irradiated in a tokamak suggests that plasma ions introduced into the coating during disruption, when the coating temperature was much higher, could not stay in the coating. Deuterium was trapped in the coating that was already cooled and crystallized, at a relatively low temperature—lower than thermal desorption maxima temperature. Moreover, based on deuterium trapping occurring at a significantly higher

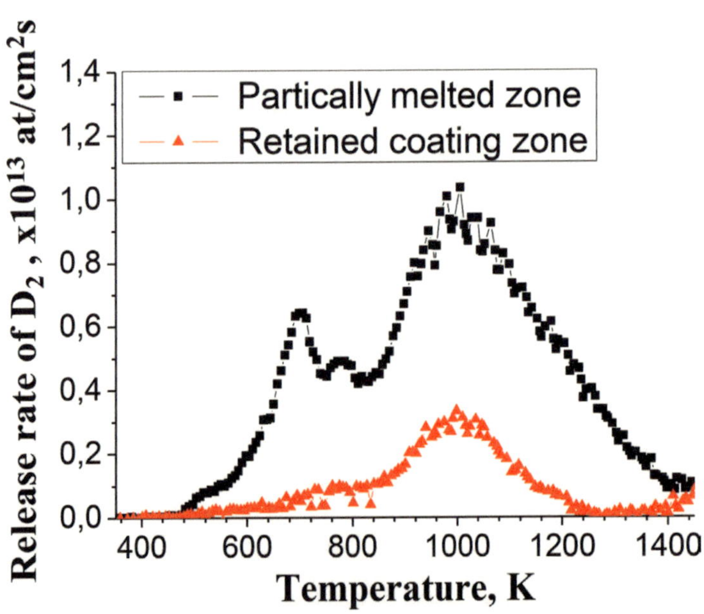

Figure 9. TDS spectra of deuterium desorbed from the sample irradiated in T-10 tokamak.

rate in an area with a partially molten coating than in an area with cracked coating, one can assume that the amount of high-temperature traps in boron carbide increased with the increase of coating's heating temperature and the loss of some boron in a film.

5. Dependence of coating morphology change on temperature and power density of irradiation

The results obtained in the experiment allowed determining the heat flux distribution of plasma irradiation on the surface of the coating to heat part I of the sample to the temperature of ≥3600 K, causing melting of tungsten, and heat part II of the sample to T ≥ 2800 K, at which the coating melts. The duration of plasma interaction with the sample was taken as 70 ms based on the measurements of the dependence of ion flux on time. Temperature calculation was done using a COMSOL Multiphysics package for two cases: irradiation from the coating side and irradiation from the substrate side. The calculations have shown that temperature difference does not exceed 100 K for both cases.

Temperature distribution on a sample surface, as well as heat loads in corresponding areas, is shown in **Figure 10**.

A graph of temperature change for coating areas 1 s after disruption based on their distance from the molten part is shown in **Figure 11**. As can be seen, maximum temperature eats the distance of approx. 5 mm away from molten tungsten which is achieved during plasma

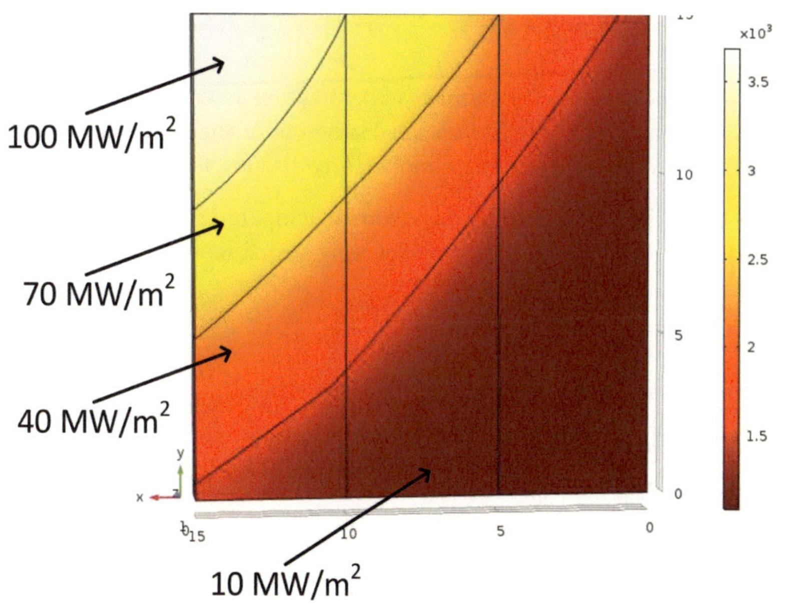

Figure 10. Maximum temperature distribution in each area and heat loads in corresponding parts of the sample (calculations). Thin lines show least distance from solidified molten tungsten for which temperature changes 1 s after irradiation are shown in Figure 11.

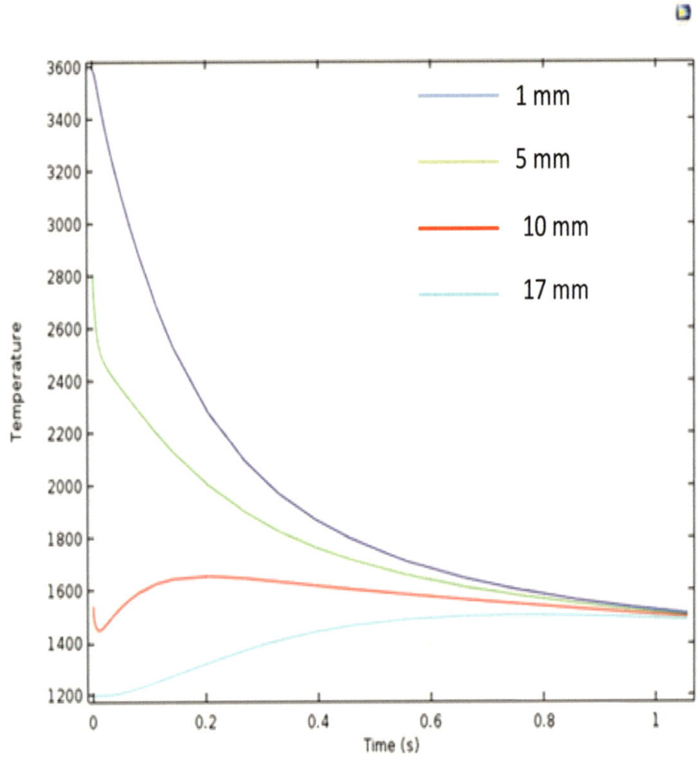

Figure 11. Temperature change dynamics of sample areas depending on their distance from molten tungsten.

irradiation. At the same time, during cooling (with radiation as a main mechanism), some heat stored in the sample spreads along its volume, leading to further increase of temperature on parts of the sample further away from the molten area 1 s after the end of irradiation.

Comparison of data allows determining approximate temperature maxima on the boundaries of each area and, as such, links the change of coating structure with temperature and power density of plasma irradiation.

Intact coating area. $T < 2000$ K.

Partially molten coating area. Temperature range 2000–2500 K.

Molten coating area. Temperature range 2500–3695 K.

Molten tungsten area. $T > 3695$ K.

6. Conclusion

The boron carbide coating deposition and testing were performed on the CODMATT facility. 5-μm-thick B_4C coating tungsten samples for experiments in T-10 tokamak were formed by atoms of boron and carbon sputtered by plasma ions from corresponding targets. The adhesion of the boron carbide coating to tungsten was tested by cycles of high-temperature hydrogen ion irradiation. During the last cycle, flux power density was elevated up to 5 MW/m^2 (E_i = 15 kV/at, W = 5.0 MW/m^2, τ = 0.6 s, t = 30 s, T = 700–1200°C). Ion irradiation did not lead to coating exfoliation, the appearance of caverns, cracks, and changes in coating composition.

The boron carbide coating on tungsten sample was subjected to intense plasma irradiation during a plasma disruption in the T-10 tokamak. Power density of an ion flow on the sample reached 100 MW/m^2, with the irradiation time being 70 ms.

The results of plasma irradiation in a tokamak can be summed up as follows:

- In the area with a temperature ranging from 2500 to 3695 K, most of the coating melts and forms globules. The structure of globules and the remaining coating layer approx. 1.8 μm thick retains boron carbide crystal lattice, but the concentration of boron is reduced to B:C = 1:1. No cracking or exfoliations are observed on this or any other areas. Part of the coating left on the surface is uniform, retaining adhesion to tungsten, which allows protection of tungsten from plasma irradiation.
- On the areas with a temperature range 2000–2500 K, partial melting is observed. The nature of melting suggests heating of individual coating areas up to high temperatures during plasma irradiation. Coating element ratio on this area approximate stoichiometric.
- At temperatures less than 2000 K, the coating is left intact on the surface of tungsten. Coating element ratio is close to stoichiometric.

The performed experiments allow making the following conclusions:

- Boron carbide coating withstands the thermal cycling and high-power density irradiation by plasma ions.
- Boron carbide coating retains uniformity and adhesion to tungsten and protects it from direct plasma interaction for temperatures up to melting point of tungsten.
- The retaining of uniform coating in contact with tungsten substrate allows renewing the coating on its surface even after high-energy plasma loads.

Author details

Leon Begrambekov[1]*, Andrey Grunin[1], Nikolay Puntakov[1], Yaroslav Sadovskiy[1], Vyacheslav Budaev[2] and Sergei Grashin[2]

*Address all correspondence to: lbb@plasma.mephi.ru

1 National Research Nuclear University (MEPhI), Moscow, Russia

2 National Research Center, Kurchatov Institute, Moscow, Russia

References

[1] Krieger K, Geier A, Gong X, Maier H, Neu R, Rohde V. ASDEX upgrade team, erosion and migration of tungsten employed at the main chamber first wall of ASDEX upgrade. Journal of Nuclear Materials. 2003;**313-316**:327-332

[2] Buzi L, de Temmerman G, Unterberg B, Reinhart M, Litnovsky A, Philipps V, et al. Influence of particle flux density and temperature on surface modifications of tungsten and deuterium retention. Journal of Nuclear Materials. 2014;**455**:316-319

[3] Krasheninnikov SI, Smirnov RD, Rudakov DL. Dust in magnetic fusion devices. Plasma Physics and Controlled Fusion. 2011;**53**(8):083001

[4] Begrambekov L, Buzhinsky O, Gordeev A, Miljaeva E, Leikin P, Shigin P. TDS investigation of hydrogen retention in graphites and carbon based materials. Physica Scripta. 2004;**T108**:72-76

[5] Maier H, Greuner H, Balden M, Böswirth B, Lindig S, Linsmeier C. Erosion behavior of actively cooled tungsten under H/He high heat flux load. Journal of Nuclear Materials. 2013;**438**:S921-S924

[6] Begrambekov LB, Buzhinskiy OI. Boron carbide properties and its use as protective coating for plasma facing components. The questions of atomic science and technique. 2016;**4**:14. [in Russian]

[7] Ayrapetov AA, Begrambekov LB, Dyachenko MY, Evsin AE, Grunin AV, Kalachev AM, et al. Stand for coating deposition and coating/materials testing. Journal of Physics: Conference Series. 2016;**700**(1):012041